Energy Economics

Energy
Economics

Helmut A. Merklein
W. Carey Hardy

Gulf Publishing Company
Book Division Houston, Texas

Energy Economics

Library of Congress
Catalog Card Number
76-46898
ISBN 0-87201-222-0

Preface

Most of the material in this book was written in the period 1973 to 1977 and thus extends back in history to the now remote pre-embargo era. In fact, the earliest writing on which this book draws was a paper presented at the Colloque sur les Hydrocarbures Gaseux et le Développement des Pays Producteurs, held on May 17-19, 1973, at the University of Dijon, France. The title of that paper (Aspects Réels et Monétaires de la Crise d'Hydrocarbures), given six months prior to the embargo, alludes to the hydrocarbon crisis, a reference to the developing shortage that was at the time well known to all energy experts, even though the U.S. public did not then, as in part it does not now, accept the legitimacy of that shortage. The great majority of the subsequent articles appeared in *World Oil,* where their enthusiastic reception gave rise to the present book.

Converting a series of articles into a book involves more than merely updating them. There was a great deal of overlapping material in the articles and every effort has been made to spare the reader the tiresome task of wading through undue repetition. However, some overlapping still occurs in places where further elimination was difficult without impairing the coherence of the text. For those occasional lapses, an apology is offered.

Being critical in nature, the book is sure to please few, if any, participants in the worldwide energy game. Many mistakes have been made in various quarters, but, overall, the United States has probably made more mistakes, and more serious mistakes, than the other big-time participants. U.S. imports of crude oil at $13.50 per barrel exceed pre-embargo imports at $3.87 per barrel by three-quarters of a million barrels per day, and domestic production has declined by 1.1 million barrels. Any way you look at it, that is not a spectacular track record.

One characteristic of energy economics is its constantly changing environment, often at great speeds which destine any book on the subject to relatively early obsolescence. For example, the publication date of this book practically coincides with President Carter's unveiling of his new energy policy. Still, much of what has been said in this book transcends time, especially the inflation calculations and the appeal for unrestricted prices in the U.S. energy sector. Chances are, complete price de-controls will never be allowed by the political sector to give the free market a fair chance to adjust the nation to a new era of high-priced energy. Thus, if the objective of this book had been to influence policy, it would certainly turn out a failure. In truth, this book has a more modest role: it calls the shots as they are seen; it is a running commentary of policies as they are and as they might have been; it is, ultimately, a resigned testimony to the weakness and ineptness of the U.S. response to a major economic problem.

Helmut A. Merklein
March, 1977

Contents

Metal Fast Breeder Reactor (LMFBR) System, 168; Heavy-Water Reactors, 173; The Nuclear Energy-Fusion Resource System, 173.

1
Energy: A Definition

Energy may be said to be the issue of this decade. It has become a subject of debate in Congress, in the White House, and in the business and academic communities. Consumer advocates have jumped on the bandwagon. The press and television feature energy specials, even the man in the street talks about it.

All this is very new. When energy was plentiful and reliably available in the United States, its existence and availability were taken for granted. For that reason, energy ranked on a par with motherhood as a conversational topic: no one ever mentioned it.

Even though many of the facts on energy are known and generally accepted, their interpretation has become a controversy in and of itself. For example, the question of currently existing proven U.S. oil and gas reserves is no longer considered controversial, except perhaps by a few individuals with axes to grind. However, the projection of future recoveries, based on the extrapolation of past U.S. exploration and production experience, is an ongoing controversy, and an honest and intellectually deserving one.

A book dealing with energy economics, and especially with energy policies, must of necessity touch on controversial issues. This book is no exception. But there is one area where all parties to the energy game agree: the definition of energy in physical terms. This chapter deals with the physical principles of energy and related concepts.

Mechanical Energy

Even though energy was used in pre-historic times (using wood as fuel for purposes of space heating, for example), the concept of energy itself, and of the physical principles behind it, had defied definition until the early nineteenth century. This is not surprising since energy, unlike matter, cannot be seen, touched or smelled; it is a concept as abstract as the mathematical concept of zero.

It has long been known that falling objects have the capacity to perform work, either constructively by driving a stake in the ground or destructively by hitting someone's toe instead. Nor did it take much imagination to determine that heavy objects perform more work than light ones. According to Aristotelian theories, that was so because heavy objects were presumed to fall faster.

This idea was shattered in the late sixteenth century by Galileo Galilei, whose curiosity was aroused by a lamp swinging on a long chain. In a series of experiments with pendulums Galileo discovered that on its downward swing a pendulum somehow picks up the capacity to climb back, or almost back, to its original height. Without formalizing the concept into a theory, Galileo was closely on the tracks of what today is called potential and kinetic energy, and their mutual convertibility. Having the Leaning Tower conveniently at hand, so the legend goes, Galileo went on to prove experimentally that two balls of different weight drop equally fast.

A British physicist, Thomas Young, first defined energy as the capacity to do work. That definition, applied in the early nineteenth century to mechanical energy, is still valid today.

A rock resting on a shelf one foot above the ground is said to have potential energy. As long as it remains undisturbed, it is incapable of performing work. However, if dropped from the shelf, its potential energy is converted into kinetic energy which, on impact, does perform work. The question is how much work or, alternatively, how much energy (potential or kinetic) does the rock have?

The potential energy of the rock is equal to the work required to raise the rock into position. In raising the rock it is necessary to overcome the gravitational force (weight) that holds it down. The greater this gravitational force, and the greater the height to which the rock is raised, the greater is the amount of work required to put the rock into its new position, and the greater the rock's potential energy.

In the British system of measurement, the so-called foot-pound-second system or f-p-s system, the potential energy of our rock is easily defined and measured since the pound (a unit of force) is one of the three elementary building blocks of the system, as is length, expressed in feet. Thus, the potential energy of a rock located at a given height h above ground is as follows:

$$E_p = f \cdot h, \tag{1}$$

where E_p = potential energy (ft-lb)
    f = gravitational force or weight of rock (lb)
    h = height (ft)

When the rock is pushed off the shelf, it is pulled downward by the earth's gravitational attraction. Immediately before impact its height is equal to, or almost equal to, zero so that its potential energy becomes zero in accordance with the preceding equation. But the rock still has energy or the capacity to do work. What happens is that the rock's potential energy is transformed into kinetic energy. Theoretically, the rock's kinetic energy just before impact (when its potential energy is zero) is exactly equal to the potential energy it used to have before the fall, at which time its kinetic energy was zero (in actuality, the kinetic energy of the rock at this point is slightly less than its previous potential energy because it has used some kinetic energy in overcoming air friction during the fall).

The kinetic energy of a falling rock is expressed by the following relation:

$$E_k = \tfrac{1}{2}\, m \cdot v^2 \tag{2}$$

where E_k = kinetic energy (ft-lb)

m = mass (slug)

v = speed (ft/sec)

The preceding equation explains why the rock's kinetic energy is zero before the fall: its speed is zero.

The metric system differs from the British system in two ways: it uses different units of measurement, and it uses different building blocks. While both systems use the notion of length (meter/foot) and time (second/second), the metric system uses mass (kilogram) as the third elementary block, as opposed to the British use of force (pound). Thus, force is a derived unit in the metric system. Its basic unit of measurement is the Newton, which is defined as that force which, if allowed to act on a mass of one kilogram, accelerates it by one meter per second^2. In practice, one newton is approximately equal to $\tfrac{1}{10}$ kilogram-weight.

The basic unit of energy in the metric system is the joule, which is the energy that must be used in raising an object weighing one newton one meter. One joule corresponds to approximately 0.74 foot-pounds. Neither the joule nor the foot-pound is generally used as a measuring rod in energy economics.

Other Forms of Energy

So far our discussion has been in terms of mechanical energy, but there are other forms of energy, such as heat, chemical, electrical, and nuclear energy, to name the most important ones. All of these types of energy are interrelated.

Chemical energy is released (or absorbed) in the form of heat when atomic bonds are broken through chemical reactions. The burning of fuel involves such a reaction; the energy so released can be measured by observing the number of degrees a unit amount of fuel can raise the temperature of a standard recipient material such as water. The heat required to raise one kilogram of water by one degree centigrade is defined as one kilocalorie. In the British system, the basic unit of heat energy is the British thermal unit or Btu, which is the heat required to raise the temperature of one pound of water by one degree Fahrenheit. One kilocalorie is approximately equal to 4 Btu's. Because most fuels release this energy in the form of heat, both the kilocalorie and the Btu are often used in energy economics as measures of energy.

Chemical reactions may also give rise to electrical energy, as in the normal car battery. The usual measuring rod of electrical energy is the kilowatt-hour used in both the British and the metric systems. A kilowatt-hour is the energy that is required to burn ten 100-watt light bulbs (= one kilowatt) for one hour. One kilowatt-hour is equivalent to 860 kilocalories or 3,413 Btu's.

Nuclear energy, used either constructively or destructively, is tapped through processes that involve the giving up of heat. The theoretical foundation of the nuclear age rests on Einstein's mass-energy equivalence, developed in 1905:

$$E = m \cdot c^2 \tag{3}$$

where E = energy (in $9 \cdot 10^{16}$ joules, if)
 m = mass (kilograms)
 c = the speed of light (approx. 300 million meter/sec).

The energy equivalence of just one pound of any substance, in accordance with Einstein's equation, is 11 billion kilowatt-hours, enough to supply the United States with electric energy for several days.

Nuclear reactions involve the transformation of mass into energy. In nuclear *fission,* the liberated energy can be calculated by Einstein's equation from the difference in mass of the original nucleus and the total mass of the fission products. The naturally occurring uranium isotope U-235 and the man-made element plutonium were the first fissionable materials used in atomic bombs (Hiroshima: U-235; Nagasaki: plutonium). Both of these elements play major roles in the nuclear reactors used for generating electric energy. Unfortunately, fissionable materials are limited in supply.

No such supply problem exists in nuclear *fusion,* such as the hydrogen fusion process, where two hydrogen isotopes, deuterium and tritium, are fused to form helium. The hydrogen fusion process has not yet been harnessed in a controlled reaction usable for peaceful purposes.

In energy economics two additional common measures are the ton of coal equivalent (TCE) and the barrel of oil equivalent (BOE). As the names suggest, these units reduce the average caloric value of a ton of coal and a barrel of oil to a standard. The energy contents of various fuels are given in Table 1-1.

Because the various types of energy are all manifestations of the same phenomenon, they are convertible into each other, more easily in some cases than in others. This convertibility makes energy particularly useful to

Table 1-1
Energy Content of Various Fuels

Fuel	Quantity	ft-lb	joule	kilocal	Btu	kWh	BOE	TCE
				Average Energy Content				
Natural Gas	1 cubic foot	7.7×10^5	1.05×10^6	252	1,000	0.30	0.0002	0.00004
Gasoline	1 gallon	9.6×10^7	1.3×10^8	3.0×10^4	1.2×10^5	36	0.021	0.0046
TNT (explosion)	1 ton	3.2×10^9	4.4×10^9	1.1×10^6	4.2×10^6	1,200	0.71	0.15
Oil	1 barrel	4.5×10^9	6.1×10^9	1.5×10^6	5.8×10^6	1,700	1	0.22
Wood	1 ton	7.4×10^9	1.0×10^{10}	2.4×10^6	9.5×10^6	2,900	1.71	0.37
Coal	1 ton	2.1×10^{10}	2.8×10^{10}	6.8×10^6	2.7×10^7	7,800	4.59	1
Uranium	1 gram	6.1×10^{10}	8.2×10^{10}	2.0×10^7	7.8×10^7	23,000	13.5	2.95
Deuterium	1 gram	1.8×10^{11}	2.4×10^{11}	5.8×10^7	2.3×10^8	66,000	38.8	8.46

mankind: it permits the generation of energy at a favorable geographic location, by means of a power dam, for example, where water is available under the right conditions, and it facilitates the subsequent transmission of energy hundreds of miles away to the end user.

An example of one such chain of useful energy conversions is the nuclear reactor, where nuclear energy is transformed into heat energy which is used to produce steam. Expansion through steam turbines converts the heat energy into mechanical (kinetic) energy which is used to drive generators and thus convert kinetic into electrical energy. Transmitted into a home, the electrical energy is then used in its end purposes of heating, lighting, or through a reconversion into kinetic energy by means of electric motors in appliances such as washing machines, dryers, electric drills, etc.

The sun is the most important nuclear reactor as far as man and his immediate environment are concerned. It generates energy through the conversion of mass in a controlled hydrogen fusion reaction—essentially the same reaction man has learned to unleash but not yet to control. Since the sun radiates its energy in all directions, and given the enormous distance of 93 million miles (on average) separating the sun from Earth, it is clear that the earth receives only an infinitesimal fraction of the sun's total energy production, two billionths of it to be exact. Yet this energy produces the winds on our planet, including hurricanes fifty times more powerful than the first atomic bomb. That same sun is responsible for supplying Earth's rivers and lakes with fresh water by a complicated cyclical process of evaporation, cloud formation, rain, drainage into the oceans, and re-evaporation. Because the sun provides the required energy for all plant growth and, therefore, for all plant and animal life on earth, conventional fuels such as coal, oil, or natural gas that have been formed from plants or animals of a past era are nothing but storage vessels of solar energy received on earth millions of years ago.

The term "generation of energy" is a misnomer. Energy cannot be generated or destroyed, it can only be converted from one form to another. An electric generator is in reality an electric convertor, transforming mechanical into electric energy. Since Einstein, the concept of the indestructibility of energy has been broadened to include mass.

The processes by which given forms of energy are converted into others have different efficiencies. For example, a gasoline engine in a car runs at about 25% efficiency. This means that only one quarter of the available heat content of the fuel is actually converted into useful kinetic energy. The rest is used to overcome friction, to heat and discharge exhaust gases, etc. The conversion efficiencies of other processes are as follows:

Coal-fired electric generator plant	40%
Steam turbine	40%
Electric motor	25%
Piston-type steam engine	10%
Solar heat collectors	4%

It was mentioned earlier that different forms of energy are often measured with different measuring rods. Mechanical energy, for example, is usually expressed in foot-pounds (British system) or in joules (metric system), while the predominant units of heat energy are the Btu (British) and the kilocalorie (metric), and so on. Because of the mutual convertibility of different forms of energy, the various units of measurement are also convertible. For example, it has been mentioned that one kilowatt-hour, which is the basic unit of electrical energy in both the British and metric systems, is equivalent to 860 kilocalories, the metric unit of heat energy. Other conversions are possible, and the applicable conversion factors are listed in Table 1-2.

As can be seen from Table 1-2, one barrel of oil has, on average, a calorific value or heat content of 1.5 million kilocalories or 5.8 million Btu's. Alternatively, one barrel of oil is the energy equivalent of 4.5 billion foot-pounds, or 6.1 billion joules, or of 1,700 kilowatt-hours, etc.

Table 1-2
Energy Conversion Factors

	ft-lb	joule	kilocal.	Btu	kWh	BOE	TCE
One ft-lb .... =	1	1.36	3.24×10^{-4}	1.29×10^{-3}	3.76×10^{-7}	2.2×10^{-10}	4.8×10^{-11}
One joule ... =	0.74	1	2.4×10^{-4}	9.5×10^{-4}	2.78×10^{-7}	1.6×10^{-10}	3.6×10^{-11}
One kilocal.. =	3088	4187	1	3.97	1.16×10^{-3}	6.7×10^{-7}	1.5×10^{-7}
One Btu ..... =	778	1055	0.25	1	2.93×10^{-4}	1.7×10^{-7}	3.7×10^{-8}
One kWh ... =	2.66×10^{6}	3.60×10^{6}	860	3413	1	5.9×10^{-4}	1.3×10^{-4}
One BOE ... =	4.5×10^{9}	6.1×10^{9}	1.5×10^{6}	5.8×10^{6}	1700	1	0.22
One TCE ... =	2.1×10^{10}	2.8×10^{10}	6.8×10^{6}	2.7×10^{7}	7800	4.59	1

The rate at which energy is applied or used is called power, which is measured in terms of energy per time unit. For example, so much energy is required to dig a three-foot ditch 100 yards in length, regardless of how the ditch is dug. However, a mechanical digging machine can dig the ditch much faster than a man equipped with a shovel, because the machine operates under greater power: its energy output per second is considerably higher than man's.

Although power can be expressed in many units, the most prevalent ones are kilowatts and horsepower. The kilowatt is 1,000 watts; one watt represents an energy rate of one joule per second. Because the watt is too small a unit for normal industrial applications, the kilowatt generally is used in its place. For large power units, such as power stations, the megawatt is sometimes used. One megawatt equals 1,000 kilowatts or one million watts. One horsepower is equal to 0.746 kilowatt. To be noted is the existence of a metric and so-called U.S. horsepower unit. However, the two units are for all practical purposes the same in magnitude.

By convention, mechanical power is generally measured in terms of horsepower; electrical power is measured in kilowatts. For example, we speak of a 300-hp engine, and not of a 224-watt engine.

It has been mentioned that energy can be stored. A car battery is such a storage device. If more energy is withdrawn from the car battery than is put back into it, as in the case of a malfunctioning alternator, the stored energy will eventually be exhausted. This is also true for the solar energy that is currently stored in the world's conventional fuels: coal, oil, and natural gas. The withdrawal rate of these energy sources greatly exceeds the regeneration rate, and this, in a nutshell, is the current energy dilemma, with or without OPEC, and it is the topic that will be dealt with in the pages that follow.

2
The
Energy
Crisis

i. International Monetary Considerations

In the Fall of 1973, the announcement and subsequent implementation of the Arab oil embargo stunned the world. The reason for the embargo was admittedly and openly political. But the power to impose such an embargo was decidedly economic. In a 1972 lecture, the author dealt specifically with the inevitable growth in power of the Organization of Petroleum Exporting Countries (OPEC), stating that this implies "a further and more forceful implementation of noncompetitive market strategies." The lecture was never published, probably because it was not in line with the then commonly held view that cartels, as anyone should know, are bound to break up under their own weight. That view has had to be reviewed in the light of subsequent events.

Whether the oil-importing nations liked the oil embargo or not, at least it was a fact that could be grasped. There was no room for controversy; the last tanker docked on the East Coast on Sunday, November 18, 1973. Whatever the cause, whatever the repercussions, whatever the ensuing emergency measures in the United States, for awhile there would be no more oil coming to the United States from Arab member countries of OPEC.

A new chapter in the history of U.S. energy was about to be written, a chapter that brought a great deal of suspense, many surprises, some discomfort and, perhaps, drama. This might have been a good time to look back and to ask where we had gone wrong. Could we have avoided the prospect of a cold winter? Had the oil industry overlooked certain possibilities? On the national level, had there been an earnest search for comprehension of the Arabs' point of view? What about our domestic full-employment policies? Have they not contributed to anti-American sentiment abroad? And if so, were foreign complaints in this regard justified? Did we ever bother to take notice? Did the international money system really discriminate in our favor, as many foreign economists had charged? Where were our economists, and why did they not offer a solution to the impending problems?

These are the questions that will be posed in this book which, for that reason, is bound to become somewhat controversial. Yet, there is merit in asking some long-neglected questions which may suggest new directions to those who sense the need for breaking new ground.

On August 15, 1971, President Nixon announced the most far-reaching change in economic policy ever devised by this country in peace time. In essence, his new economic policy contained three major elements:

1. A freezing of wages and prices that was later followed by controls;
2. A fiscal package featuring various tax concessions and a reduction in government spending;
3. A "temporary" suspension of the convertibility of U.S. dollars into gold.

The last item, suspension of convertibility, was least noticed—for a good reason: it did not directly affect any U.S. citizen for whom the gold convertibility of dollars had been suspended in 1933, during the Great Depression. Yet, in the long run, each and every U.S. citizen was bound to feel the repercussions of this policy move. A discussion of the historical development of the current international money crisis is indispensible for a thorough understanding of the world oil situation.

The days of political or economic isolation are forever gone. Nations engage in international trade and in international capital investment; there is foreign aid, and various nations pursue military assistance programs. In short, the productive process carried on by mankind has assumed global dimensions. Necessarily, this rapid increase in international economic activities has required that various nations deal with each other's currencies. World trade alone in 1973 amounted to 450 billion U.S. dollars, which corresponds to the total output the United States is presently capable of producing in a period of 4½ months.

Bretton Woods, IMF, and the Value of the U.S. Dollar

To deal with the expected post-war growth in international economic activities, and indeed to facilitate it, a set of rules was laid down at an international monetary conference in July 1944. This is the well-known Bretton

Woods Agreement that led to the creation of the International Monetary Fund, or IMF. Among the stated objectives of the IMF were:

1. Expansion and balanced growth of international trade
2. Promotion of exchange stability
3. To shorten the duration and lessen the degree of disequilibrium in the international balances of payments of member countries.

The only one of the three preceding objectives that was in fact met was the growth of international trade. Even though the Bretton Woods Agreement is credited with this achievement, there is considerable room for doubt concerning such a contention. This will be discussed later in more detail in connection with floating exchange rates, Chapter 4.

There is no doubt that the other two objectives were completely missed. For example, from 1951 to 1971, more than 100 currencies were devalued or revalued, and more than 15% of these currencies have been devalued at least twice.[1] Even major currencies that were supposed to set an example for all others were subject to occasional devaluations. Witness the November 1967 devaluation of the British pound and the August 1969 devaluation of the French franc.

The performance record in regard to objective (3) was even worse. By definition, the United States experiences a balance-of-payments deficit when more U.S. dollars leave the country than enter it. Dollars leave the country when Americans buy goods and services abroad (import), or when they invest abroad. Conversely, dollars enter the country when foreigners buy U.S.-made goods and services or when foreigners invest here. These are not the only causes of international dollar flows, but they are the major ones. Others, such as international remittances and pensions, military and non-military grants, etc., are of some importance. Yet, having mentioned them, things can be simplified by discussing the U.S. balance-of-payments deficit on the basis of trade-induced dollar payments and of capital flows.

The Bretton Woods Agreement worked reasonably well in the first few years of its existence, perhaps because the IMF, being in its infancy, moved with caution. By 1950, the first signs of trouble appeared. From then on, until the demise of the system, the United States persisted in accumulating balance-of-payments deficits, with only rare and insignificant exceptions. In fact, in the period 1958-1973, the United States did not once have a positive balance of payments, except for a truly minuscule $171 million surplus in 1968. This means a practically continuous balance-of-payments deficit over a period of 16 years. Insofar as the United States is concerned, the IMF failed to meet the balance-of-payments objective.

By mid-1971, the United States had accumulated a balance-of-payments deficit of some $53 billion. The country had, as it turns out, achieved a position of immunity against balance-of-payments deficits. Or so it was thought, when, on August 15, 1971, the U.S. dollar caved in.

How, precisely, did the United States get away with this for so long when no other nation could? Simple: by convincing the creditor nations to hold the U.S. dollar itself as a means of settling its deficit. And, in fact, why should they *not* want to hold dollars, since the U.S. government continually assured everyone that the dollar was as good as gold? The fact that it had only some $10.5 billion worth of gold in reserve appeared to be no cause for concern. Of course, this made a fractional reserve system out of the international U.S. dollar market, but almost all countries today operate *internally* on such a system, and familiarity breeds contempt.

That there was a vital difference between an internal and an international fractional reserve system, one of differing identities of interest, was not noted until it was too late. (This point will be discussed later.) As U.S. dollars continued to pile up in foreign central banks, it was always thought that a simple reversal of policies could, in due time, reverse the balance-of-payments trend.

In the meantime, as long as the U.S. government adhered to its policy of keeping its dollars pegged to gold and convertible into gold, a run on the U.S. gold stock was not likely. This was all the more true, since the United States creditors were in the same boat with the U.S. government. If a run were to be made, the United States would lose—but so would most every free nation, since they would be left holding U.S. monetary units for which there was no gold backing. In time it became clear that the Bretton Woods Agreement discriminated, in practice, in favor of the United States. Of course, that had not been its intent, but it was the effect. There was something inherently unjust in the system. This is made clear by the following example.

Assume private citizen Joe Blow is given the monopoly of providing the United States with its required money stock. As this nation's GNP grows, the volume of its transactions will grow, and more money will be needed. Short of throwing the new money out his window, Mr. Blow will put it in circulation through purchases on his personal accounts or by loaning it out. He will literally spend himself rich. That is, the act of supplying the United States with its needed money will be a source of wealth to the supplier.

Indeed, monetary authorities in a growing country show a profit simply by supplying the money that is compatible with continued growth. Domestically, this is not disturbing. After all, the U.S. Treasury (which ultimately gets these profits, even though they are technically made by the Federal Reserve System) represents all Americans, since it is part of the U.S. government. No person or group derives an advantage from these profits.

Internationally, the picture is different. Again, international money is provided to assure continued growth of world trade. And again, the monopolist providing the money makes a profit. But this time the profit-making monopolist and the market served do not share identical interests.

Under the Bretton Woods system, the United States had assumed the role of the monopolistic money-supplier of the world. Just as in Mr. Blow's case, the U.S. dollar was put in circulation throughout the world by purchasing goods or by making loans. In either case the monopolist stood to reap tremendous benefits. The purchase mechanism provided Americans with foreign-made goods or services, the loan mechanism (compounded by the Euro-dollar market, to be sure) permitted American corporations to make capital investments abroad, i.e., to buy up foreign productive facilities or to create such facilities on foreign soil. In return for this accumulation of wealth, the U.S. had done no more than run the money printing presses a little faster.

Whether Americans like it or not, the world at large does *not* identify with them, and any profit that accrued to them by virtue of their supplying the world with liquid funds was likely to be resented, even though such a profit can accrue to them only through the permissiveness, the indifference, or the inattention of the participating foreign nations. The only way America can make U.S. dollars available to the outside world is by incurring balance-of-payments deficits. Again, this abstracts from the multiplier effect of the Euro-dollar market and also from the creation of bookkeeping dollars abroad involving U.S. dollar accounts held abroad in U.S. affiliated or correspondent banks. The greater the deficit, the greater the international accumulation of U.S. dollar holdings, and the greater the international liquidity. But international liquidity was believed to be badly needed worldwide. One is almost tempted to conclude that it was America's duty to provide that liquidity, to fulfill its obligation to the world and spend, spend, spend.

Balance-of-Payments Deficits and the U.S. Money Supply: Exporting Inflation

The fact that the pre-1971 international monetary system favored the United States was not the only area of friction among nations. Even more annoying to our foreign trade partners was this business of exporting inflation. This, in any event, was one of the side-effects of "printing money" on a seemingly unlimited scale for foreign use.

Money, at least the so-called M_1-concept, is defined as all checking accounts plus currency in the hands of the U.S. non-bank public. That, among other monetary aggregates, is the quantity the Federal Reserve System or the "Fed," as it is commonly called, feels obligated to watch with an eagle eye. But since U.S. dollars held abroad, either privately or by foreign central banks, do not fall into this category, the Fed does not concern itself (at least not in practice) with the supply of dollars held abroad.

As an example, suppose the U.S. money supply is $200 billion and that in a given year some $5 billion wind up abroad through that year's settlement on the balance-of-payments deficit. That would leave $195 billion at home, not enough to sustain the GNP at the previous level, since a reduction in the nation's money supply de-stimulates the nation's aggregate demand. So America has what looks like a demand-induced recession on its hands.

If left alone, this would tend to redress the balance-of-payments disequilibrium. But, of course, such a condition will not be allowed to exist for long. The Fed, in this case, will count and recount its domestic money supply (M_1), find it short by $5 billion, and promptly fill the gap. This will restimulate the nation's aggregate demand. The GNP will rise. Unemployment will be reduced. The problem is solved.

Or is it? If anybody cared (or dared) to ask the one relevant question, potential problems would immediately present themselves. The question: If America is short $5 billion, where did the money go, and what is its impact at the new location? The answer: It went abroad and set up inflationary pressures there.

Preposterous? Let the record speak for itself. The first quarter of 1971 was marked by a balance-of-payments deficit (net liquidity balance which includes all trade and capital flows, except liquid private capital flows such as "hot" or speculative money) in the amount of $2.5 billion. Did the U.S. money supply decline by a like amount? Far from it, it was *increased* by $4.9 billion. A similar pattern develops quarter after quarter, as can be seen in Table 2-1 (all $-figures in billions).

Thus, by not filling the money gap created by a transfer of dollars abroad, a recession is generated at home. By filling the gap, the recession is averted, but an inflation is created abroad. What would you do? Inflate abroad, of course. It's simple, isn't it?

This sounds incredible. Surely the United States would not deliberately foster inflations abroad in order to achieve full employment at home. Or would it? Listen to the comments on the other side of the Atlantic:

1. Raymond Barre, vice president of the Commission of the European Communities warned, in April 1971, that if the massive payments deficit persists, dollar balances being accumulated by foreign central banks would sooner or later reach a limit.

2. Pierre-Paul Schweitzer, then chairman and managing director of the IMF, just before the May 3-5, 1971, attack on the U.S. dollar (the last one, by the way, the U.S. survived unscathed) stepped up his pressure on the United States to alter economic policies. For this, he fell in disgrace with the United States, which ultimately cost him his job.

3. As early as November 1970, another spokesman for the European Community made the feelings of foreign central bankers very clear:

Table 2-1
U.S. Balance-of-Payments Deficits vs. Money Supply
(1971-1974)

Year	Quarter	Balance-of-Payment Deficit (Net Liq.)	Increase in Money Stock (M_1)
1971	1	$ 2.5	$ 4.9
	2	5.8	5.8
	3	9.3	2.1
	4	4.3	0.6
	'71 Total	21.9	13.4
1972	1	3.1	5.3
	2	2.2	3.7
	3	4.5	5.0
	4	3.8	5.4
	'72 Total	13.6	19.4
1973	1	6.7	2.4
	2	1.6	7.4
	3	1.6	(0.1)
	4	1.0	5.0
	'73 Total	10.9	14.7
1974	1	1.0	3.7
	2	6.2	4.5
	3	3.1	1.1
	4	7.7	3.6
	'74 Total	18.0	12.9

Source: Federal Reserve Bulletins

"Washington's efforts to achieve full employment by 1972 would mean . . . a large balance-of-payments deficit. It would be difficult for European Monetary Authorities to explain to their public that they were accumulating dollars without limit."

In view of these problems, one would expect serious attempts on the part of the U.S. government to rectify the situation. But no. Then Vice President Agnew, speaking in early May 1971, declared that "the U.S. government will not put its economy through the wringer to deal with a temporary international situation." Treasury Secretary John Connally voiced similar thoughts: "The turbulence in international currency markets won't cause any change in U.S. policy." And finally, George P. Shultz, then Director of the Office of Management and Budget, declared that "some people seem to think we should virtually place the balance of payments at the top level of priorities . . . that is an attitude that doesn't seem acceptable to me."

And so it went. America refused to take action, dollars continued to accumulate abroad, and the inevitable finally took place. During the second week of August 1971, another run on U.S. gold was nipped in the bud when, on August 15, President Nixon suspended its convertibility into gold. America had, in fact, capitulated.

There had been previous attacks on the U.S. dollar, of course. In 1965, France converted just under $1 billion into gold. No reasons were given for this flight from the U.S. dollar and accumulation of gold, but authoritative sources confirm that it wasn't for dental purposes. The dollar came under attack again in 1968. The two-tier gold market was then created as an emergency measure: the United States no longer stood ready to exchange an ounce of gold for $35 to just anyone. Only a foreign government or central bank continued to qualify for the trade. The United States, in other words, cut itself loose from the private *international* gold market, just as it had abandoned and indeed outlawed the private *domestic* gold market in 1933.

The trouble is, when that door was shut, a small and devious backdoor was left ajar. Under the existing IMF rules, member governments, and that means practically all free world nations (the notable exception being Switzerland), were obligated never to let their money parity deviate by more than 1% up or down from the IMF-set fixed exchange rate relative to the U.S. dollar. If, under these circumstances, foreign or U.S. holders of dollars became nervous about their dollars' value, they would sell them. But if everyone wants to sell dollars in, say, West Germany and no one wants to buy them, the price of the dollar has to give. In the absence of government intervention, that price will fall and the dollar will decline in value relative to the Deutsche mark.

Here is where the IMF-fixed exchange rate came to haunt the foreign governments, in the present instance West Germany. That country was prohibited, by agreement, to let the price of the U.S. dollar fall. To prevent such a price decline, there was only one thing the West German government could do—it had to buy up U.S. dollars at the fixed exchange rate as fast as they were offered by individuals or corporations. Now the dollars were in the hands of foreign central banks, and in the days preceding President Nixon's new economic policy, that meant that these central bank-held dollars now had access to Fort Knox. But the foreign central banks were assumed to be reasonable institutions that would not make a run on the U.S. gold.

In early May 1971, a private attack on the U.S. dollar forced four countries (Germany, Holland, Belgium, and Switzerland) to buy up something like $3 billion U.S. within a few days. Finally, these and other countries decided they could support the U.S. dollar no longer and shut down all international money markets for several days. They had to, because a wholesale conversion, in Europe, of U.S. dollars into European currencies would have flooded the dollar-supporting countries with their own currencies which were neither wanted nor needed. This is the case of an instant inflation on a gigantic scale, "made in U.S.A." and delivered in Europe.

Of course, the dollar leak was not the only source of inflationary pressures in Europe. Far from it. The Europeans had read Lord Keynes, and they, too, had applied his easy-money policies with abandon. But that was their problem and of little concern to the United States.

Thus, on August 15, 1971, President Nixon, during the early signs of a renewed attack, suspended convertibility of the U.S. dollar into gold. Significantly, in his policy speech, he singled out the "international money speculators" as the one group that greatly contributed to the recurring money crises. The reference is unfortunate. Speculators do not set up pressures, they react to pressures. Speculators are incapable of destabilizing markets; they thrive in destabilized markets, however, and they are the very force that tends to stabilize them.

There was disruptive speculation that acted as a destabilizing mechanism: its seat was Washington, where it was speculated, not without a good measure of naïveté, that no speculation would take place in the presence of a massive oversupply of U.S. dollars.

The pressure that caused the panic-stricken flight out of the U.S. dollar simply reflected that the market considered the U.S. dollar highly overvalued. If the administration insisted that the dollar was not overvalued and if—in part triggered by "speculative" money transfers, but ultimately caused by persistent balance-of-payments deficits—that same dollar was forced to devalue, such a forcing of the U.S. government's hand will obviously be viewed as destabilizing by that government.

The immediate result of suspending convertibility of the U.S. dollar into gold was a joint upward floating of most major world currencies, in relation to the U.S. dollar. This, of course, meant that the dollar had been devalued, from outside, so to speak.

It was exasperating to observe our government officials insist that the dollar had not been devalued. Everybody knew that it had, and nothing was gained by this childish insistence, except perhaps that the U.S. government's monetary credibility gap was widened both at home and abroad.

By December 1971, the United States finally admitted officially what had been fact for four months. A new monetary agreement, the so-called Smithsonian Agreement, was reached among the IMF-nations. President Nixon hailed it as the greatest innovation in the history of international money. Actually, there wasn't anything really new in the agreement. The U.S. dollar was devalued by increasing the official price of gold from $35 to $38 an ounce. Even in this dark hour, the United States put on a display of strength totally incompatible with the grim hopelessness of the situation.

While devaluation was agreed on in mid-December 1971, the U.S. government was in no hurry to put it into effect; indeed, it did *not* put it into effect until May 8, 1972.

Another "innovation" in the Smithsonian Agreement permitted the IMF countries to let their currencies fluctuate 2¼% up or down from official parity, compared to the previously established 1% bracket. The move was designed to provide greater flexibility, but as far as the United States was concerned, this was totally academic. In view of its ongoing overvaluation, the U.S. dollar promptly fell to the floor and stayed there, totally indifferent to the government holding its breath.

The naïveté of the international monetary agreement of December 1971 is illustrated by the following example. Let a gallon of milk be valued at $1.00 in the market. Suppose that price controls are initiated, freezing the milk at the "market price," plus or minus 1%. Now let the country engage in a liberal money-printing policy that, in the absence of controls, would cause the price of milk to double. Obviously, under these circumstances, the controlled price of milk would rise to its legal ceiling of $1.01 per gallon. The milk, in other words, is sorely undervalued, or the U.S. dollar overvalued.

Now let the government call for "greater flexibility." Let them allow the price to rise by 10% to $1.10 per gallon (a 10% devaluation of the dollar), and then let them widen the price bracket to 2¼% either side of the fixed price. The result is clear: the price of milk will immediately rise to its ceiling of $1.13 and never budge from there. Greater flexibility? On paper, yes. In practice, no. This is what the great innovation of the December 1971 currency agreement amounted to.

In February 1973, a second devaluation of the U.S. dollar occurred. Shortly thereafter, most major countries followed the lead of Canada, Germany, Holland and others by allowing their currencies to float against the U.S. dollar. To the great surprise of those who had predicted an immediate collapse of international trade should the "chaotic" conditions of floating exchange rates ever prevail, world trade continued to prosper. Indeed, its growth rate accelerated after (which is not the same thing as "because of") the introduction of floating exchange rates, thus casting doubt on the previously mentioned assertion that the post-war growth in world trade was the result of fixed exchange rates.

Resourceful as always, the private sector responded by opening up a futures market for foreign currencies. Ironically, this so-called International Monetary Market of the Chicago Mercantile Exchange invites speculation, because it recognizes (as the government does not) that speculation *contributes* to stability.

Finally, in November 1973, the major countries agreed to demonetize gold. The word "demonetize" was not used, but the agreement allowed participating governments to sell gold on the free market; thus governments were no longer required to hold gold in reserve for the purpose of settling balance-of-payments deficits.

Dollar devaluations are costly to foreign holders of dollars. For example, the Arab oil-producing countries held about $13 billion in gross foreign liquid assets by the end of 1972.[2] A conservative estimate of 50% of this sum being held in the form of U.S. dollars (mostly in the Euro-dollar market) means that the Arab countries must have lost something like $1.3 billion in the two dollar devaluations of December 17, 1971, and February 13, 1973. No wonder they, like any holder of sizeable dollar-denominated liquid funds (banks, multi-national corporations), will get the jitters at the slightest sign of monetary nervousness, and they will react by dumping dollars. Protection is the descriptive term for such a policy, not speculation.

Actually, the U.S. government was very slow in realizing the balance-of-payments problem that the importation of oil would ultimately represent. In 1970, for example, some 3 million bpd of crude were imported into the U.S., causing an energy-related balance-of-trade deficit of some $3 billion for that year. By 1973, immediately prior to the Arab embargo, the daily oil-import had doubled to 6 million bpd. The balance-of-trade deficit would have more than doubled, since oil prices had been rising even then.

The United States needed more and more oil. One pre-embargo projection had estimated that the U.S. oil imports in 1985 would reach 15.2 million barrels per day.[3] Most of the increase would have come from the Middle East. But the OPEC countries, once burned, were increasingly hesitant to accept U.S. dollars in payment for oil. Indeed, their insistence that protective clauses against dollar devaluations be built into the sales price of oil seemed thoroughly justified on the basis of their previous experience.

It is clear that international oil cannot be meaningfully debated without taking into critical consideration the international monetary system. Yet this system is rarely included in the public energy debate. Why not? The international oil corporations have no direct policy voice in international money. This may be the reason for their reticence. But the various governments they deal with are participant policy makers in international money. These governments should be concerned, and they are. Because their foreign partners are officials in the international monetary business, the oil corporations should also be concerned, and maybe they are, privately.

Slowest of all in coming to grips with the international monetary implications of world oil trade was the U.S. government. In February 1970 the Cabinet Task Force on Oil Import Control issued its report entitled "The Oil Import Question." If there ever was an official governmental task force, this was it. No less than half a dozen cabinet members were on the task force, plus another half-dozen or so very highly placed observers.

What did this august body have to say about the international monetary system? Nothing, absolutely not a thing. It ignored its existence and relevance to the question at hand, even though the original presidential mandate called for a *comprehensive* review of the mandatory oil import restrictions that had been in effect for slightly more than a decade.

There was a discussion on the balance of payments, to be sure. What was missing was a consideration of balance-of-payments-induced repercussions on the strength of the American dollar. The potential collapse of the international monetary system, so clear to most informed observers at the time, was blithely ignored.

Considering balance-of-payments, the report, read in retrospect, is a mockery of reality. It was asserted that the elimination of import controls would lower domestic oil prices to $2 per barrel (page 39). On the subject that "the exporting countries might form an effective cartel that would charge us a monopoly price," the Task Force adopted M.A. Adelman's thesis: "That seems unlikely" (page 35). Finally, in regard to the overall impact of oil imports on the balance-of-payments situation, an unrealistically low figure (even for 1970) was given: $.3 to $.45 billion per year for each additional million barrels per day imported, with no allowance for future increases.

The scope of this chapter imposes brevity. If the international monetary system has not been covered exhaustively, it is at least hoped that this discussion has shown that not all is well in this area. See Chapter 4 for a detailed discussion of international money under the fixed exchange rate system of Bretton Woods and under the floating exchange rate system.

No one needs to be reminded that the U.S. domestic economy is in a state of disarray and with it the oil industry. The next section in this chapter takes a closer look at domestic petroleum.

ii. The Domestic Petroleum Situation

The first part of this chapter dealt with international monetary problems and their relation to oil. It was shown that the U.S. government has for decades been printing fake money for use abroad, while pursuing the only economic policy it knew well: the buying of time. This international policy is paralleled by an equivalent domestic easy-money policy.

This is not the time to deal in detail with mistaken domestic economic policies, the results of which are here for all to see. But it should be pointed out there is one significant difference between international and domestic fake money. OPEC countries, as noted earlier, cannot and will not be compelled to accept overvalued dollars. Unfortunately, U.S. citizens are not so lucky. They are required by law to accept these dollars as legal tender, and it hurts.

Competition, Supply and Demand

In discussing the domestic energy problem, a simple competitive demand and supply model will be used as a starting point to analyze the domestic output-pricing relation. Judging from its straight-jacket policies, the government either has no conception of supply and demand or else it has no faith in the free interplay of demand and supply forces.

The petroleum literature also has been using the term "demand" and "supply" somewhat loosely. For example, total U.S. oil consumption is usually called U.S. demand; conversely, that same variable (total consumption), if broken down by sources, is often called supply. The fact is total oil consumption (or total oil sales) is neither the U.S. demand nor the supply, but it is determined by the interplay of both demand and supply.

The ideal case of perfect competition is subject to exacting requirements; so exacting, in fact, that perfect competition is never attained. Nevertheless, it does provide a convenient starting point for this discussion.

In essence, there are four basic requirements for perfectly competitive markets:

1. There must be large numbers of producers, each one sufficiently small relative to the market so that he cannot individually affect that market. The petroleum industry goes a long way toward meeting that requirement: there are several thousand oil and gas producers in the U.S., the largest of whom accounts for less than 8% of domestic oil production. Similarly, the largest U.S. refiner has less than 9% of the total U.S. refining capacity.

2. The one requirement most nearly met in the petroleum industry is product homogeneity. There are quality differences for different types of crude oil, such as different API gravities or sulfur contents, but these differences are minor. Crude oil is freely mixed; in fact, it is sometimes deliberately (and quite legally) mixed on location to upgrade one horizon's quality. An example of a liquid that does not possess any homogeneity is wine. A bottle of Chateau de Cardaillan cannot (must not) be mixed with a bottle of Liebfraumilch. And, indeed, the wine market is highly monopolistic, at least insofar as individual wines are concerned. However, easy substitution of one wine for another by the consumer makes for a very competitive overall wine market, regardless of the individual monopoly status on the production side. This is a point that also applies to crude oil and that, for this reason, deserves close consideration by legislators.

3. Free mobility of resources is the one requirement that the petroleum industry does not even remotely meet. The reason is technical, not economic, but the effect is economic in the sense that it has a bearing on the industry's competitive structure, as well as on its vulnerability.

As far as its labor force is concerned, the petroleum industry is subject to more or less the same institutional environment as any other industry comparable in size. But this is not the case with the resource called capital.

Capital can be moved into the petroleum industry as easily as into any other industry. It is no more difficult to build another refinery or to drill another well than it is to build another manufacturing plant. But capital is notoriously difficult to move out. A bicycle manufacturing plant, to use an example, can be converted to build motorcycles or some other durable good. A refinery has no second alternative, nor does the oil well. A dry well represents an irrevocable loss; an idle plant does not.

4. Perfect knowledge of prices, costs, output rates, etc., is the fourth requirement for perfectly competitive markets. Again, the petroleum industry is no better or worse off than most other industries, especially now that it is government controlled.

Demand and supply models, as used in the literature of applied economics, are usually of the perfect-competition variety and are subject to the four requirements listed above. This deserves mention, since any variation from these requirements affects the market structure and thus the model itself.

A competitive demand-and-supply model is shown in Figure 2-1 (prices and output rates in this and following figures are hypothetical). The demand curve reflects output rates U.S. buyers are willing and able to absorb at given prices. Or, and this is the same thing, the demand curve reflects maximum prices that consumers are willing to pay for given output rates. As will be seen later, both definitions are equally important.

The supply curve shows output rates producers are willing and able to maintain at given prices. Alternatively, the supply curve shows the minimum prices that must be paid to producers to get them to produce given output rates. Incorporated in the supply curve is the competitive rate of return.

Thus, the demand curve reflects the buyer's preferences and purchasing power. There is no need to belabor the fact that the demand curve is negatively sloped. A cut in prices, everything else remaining the same, will increase the quantity of goods people are willing to buy. In economic jargon, a reduction of prices will increase the quantity demanded.

The supply curve reflects conditions as they arise on the producers' side. The level of technology, capital intensity, and degree of monopoly are examples of factors having a bearing on an industry's supply curve. The petroleum industry represents a textbook example of a positively sloped supply curve, indicative of rising unit costs as a result of increased output rates. To step up the producing capacity of this nation, more and deeper wells will have to be drilled in less favorable locations. Translated into dollars, this means higher unit prices.

As can be seen Figure 2-1, oil price and output rate are codetermined by the demand for and supply of oil. Such an equilibrium price-output position is given by Point A, where demand and supply curves intersect. Thus, in terms of Figure 2-1, if the price of oil is $3.50/bbl., producers are willing to produce 10 million bpd, which happens to be the exact quantity buyers are willing to take off the market at that price. The market is said to be cleared at a price of $3.50, and Point A is called the market-clearing equilibrium position. This position is stable.

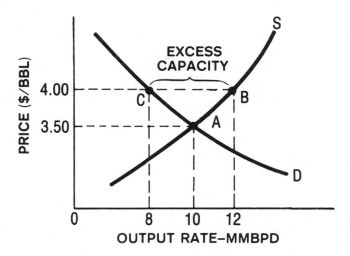

Figure 2-1. *Competitive demand and supply.*

To prove the stability of Point A, a temporary price of $4 may be considered. This is a competitive market, and a move away from Point A is difficult to conceive, precisely because the position is stable. A small trick is required, and that trick will be the assumption that the producers forget over the weekend what the market price was the previous week, and they all charge $4/bbl. on Monday. As it turns out in Figure 2-1, at that price the producers would want to increase their output rates to 12 million bpd, Point B on the industry's supply curve. (Again, this assumes competitive markets. The situation is different in monopolistic markets, where producers might be content with reduced output rates at abnormally high prices.)

However, to increase production to 12 million bpd would be a mistake because, at a price of $4/bbl., buyers would only take 8 million bpd off the market, Point C on the oil demand curve. Thus, there would be an excess supply of 4 million bpd at the greater-than-equilibrium price of $4. The industry would be faced with excess capacities and growing inventories, and it would not be long before pressures would be felt in the industry to reduce inventory levels by cutting prices.

The downward adjustment of crude oil prices will discourage production, causing the producer to move down along the supply curve toward Point A, and it will stimulate buying, causing purchasers to move down along their demand curve toward Point A. Where the two curves meet, the excess supply is removed, the market is cleared, and pressure to cut prices is eliminated (Point A).

A similar process can be used to show that a lower-than-equilibrium price will return the market to the equilibrium position designated by Point A. The only difference is that it is the consumers (rather than the producers) who will bid up prices in response to shortages that result from low prices.

Tax Incentives

The market-clearing position (Point A) is stable, and the economy will be producing 10 million bpd at a price of $3.50/bbl. This output rate will henceforth be called *total sales*. Unfortunately, Figure 2-1 depicts a model that is much too simple to be useful. For example, no consideration is given to the tax-levying government sector which, manipulating the level of taxation, affects the petroleum industry's supply curve.

The power of taxation allows the government to selectively stimulate or hold back growth of different industries. For example, many industries are engaged in pure research (as opposed to commercial research) and operate as nonprofit organizations exempt from any taxation. Such an arrangement stimulates the growth of these industries, which is precisely what the government wants. More cancer research is conducted under tax shelters of this kind than would be conducted in their absence. Conversely, cigarettes and alcohol are goods subject to special taxation. This raises the price of these goods to the consumer, which tends to reduce total sales and restrict growth.

All U.S. firms enjoy the tax incentive called depreciation, and they are all presently subject to the investment tax credit. Moreover, the petroleum industry can expense certain portions of its drilling costs, a sort of instant depreciation. All of these tax incentives stimulate the output and growth of the U.S. petroleum industry by lowering the price of crude and, consequently, of crude oil products. That, in fact, is their intent.

Contrary to popular belief, tax incentives *do not* boost profits. This is true empirically, since the oil industry has had these incentives (plus the now-defunct depletion allowance) for decades, yet profits have not been abnormally high (especially domestic profits), and they were depressed in 1971-1972.

Nor does economic theory claim that tax incentives provide permanent excess profits. Suppose the government decides to subsidize hamburgers at the rate of 20 cents each, without further controls. The immediate impact of such a program would be that the hamburger stands would make a killing in the market by simply pocketing the subsidy. Such a get-rich-quick scheme would not go unnoticed, and new hamburger stands would open up on every corner. To sell their hamburgers, these stands would compete on prices, and the established stands would have to follow suit. New entry and continued price competition will erode away the "windfall profit." Indeed, only when the resulting profit rate has reached the pre-subsidy rate will further entry and price-cutting stop. The long-term result, then, of such a policy will be more hamburger stands, i.e., more hamburgers at lower prices, with profit rates the same as before.

Translated to petroleum: The tax incentives have done nothing to boost profits, but they have contributed to abnormally high consumption of crude oil at below-free-market prices. "Misallocation of resources" is the technical term for this phenomenon.

Of course, when tax incentives are introduced or eliminated, raised or lowered, there will be transitory profit repercussions. This is why the oil industry was opposed to the elimination of the depletion allowance. During the period of readjustment, profits were expected to be adversely affected.

The effect of tax incentives upon a demand-and-supply model is shown in Figure 2-2. Two supply curves exist now. Supply curve S_{soc} is the competitive supply curve from Figure 2-1. It depicts the production cost to society. No matter what the government's form or level of taxation, the S_{soc} curve depicts minimum unit prices required (given the present level of technology, etc.) to elicit given production rates, i.e., the prices society will ultimately have to pay to get the oil, either directly through prices or indirectly through taxes.

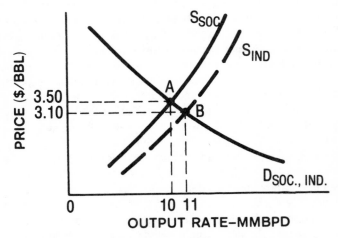

Figure 2-2. The effect of tax incentives on demand and supply.

To the industry and the consumer, however, the dashed curve S_{Ind} is now relevant: this curve depicts industry cost conditions. The vertical difference between the two curves is the per barrel tax rebate made available to industry. Since the top supply curve represents actual costs and the bottom curve the industry's costs, somebody else must absorb the difference. As mentioned previously, that somebody else is the government and, ultimately, the taxpayer.

The result of such a downward shift of the industry's supply curve is twofold: the market price of oil is reduced (from $3.50 to $3.10 in Figure 2-2) and total output of oil is raised (from 10 to 11 million bpd, Point B). Such a stimulation of oil production is, in fact, the government's reason for having the tax incentives in the first place.

There is no question that the U.S. oil industry would be less significant in size if the depletion allowance and the expensing mechanism had never existed.

Price Regulation vs. A Free Market

The model shown in Figure 2-2 is static in that it depicts the nation's supply-and-demand conditions at a given point in time. Oil supply and demand both vary over time. In the United States the supply of oil is declining and the demand, rising. A decline in oil supply in the face of a given demand will drive up prices and reduce output. This case will be taken up later. The case of an increasing demand for oil and the resulting import problems are the next topics to be discussed.

In the absence of foreign supply markets, an increase in demand will cause both the price of oil and the output rate to rise. This is illustrated in Figure 2-3, where the given industry supply curve and initial demand curve are reproduced from Figure 2-2. An increase in demand from D_1 to D_2 shifts the nation's market-clearing positon from Point B to C. The price rises from $3.10 to $3.30/bbl., and the total sales are increased from 11 to 13 million

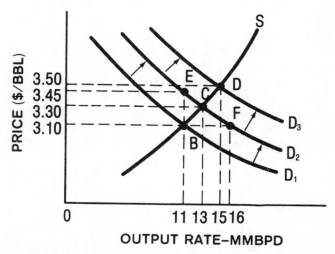

Figure 2-3. *The effect of increasing demand on domestic oil prices and output rates.*

bpd. A further increase in demand (D_3) pushes up the price and sales of oil even more (Point D).

Thus, in the absence of foreign markets, and barring government price regulations, increasing domestic demand exerts a steady and irrevocable upward pressure on the price of oil.

A detailed look at the adjustment mechanism is in order. The initial market-clearing point will be assumed to be Point B. An increase in demand (from D_1 to D_2) immediately establishes a shortage of oil: at the old equilibrium price of $3.10 per barrel, the people now wish to buy 16 million bpd (Point F). The supply at that price remains 11 million barrels, so the shortage is equal to 5 million bpd.

In discussing the stability of the market-clearing point a volume adjustment mechanism was used: the price was varied and the resulting changes in volumes supplied and demanded were observed. For reasons that soon will be clear, the equivalent price adjustment mechanism will be described at this time.

Given the 5 million bpd shortage as stipulated above (Figure 2-3: Points B and F), consumers are competing for the current output of 11 million bpd. In fact, demand curve D_2 shows that they are willing to pay $3.45 per barrel (Point E). In the absence of controls they will be paying that much, and oil producers are going to make a "windfall profit" (the term used by the government) or an economic rent (the economic term). That is, at Point E, the oil companies will be receiving an amount of $.35 per barrel over and above their cost of production plus a competitive rate of return, as given by the supply curve. This is the very incentive that is needed and which the free market provides automatically to assure the required expansion.

There are two morals to the story. First, it is clear by now that price ceilings are counterproductive because they deny the incentives needed for expansion. Inevitably, if demand rises, price ceilings create shortages. Witness the creation of shortages of all kinds of goods and services in the period 1971-1974, when price controls were used in the United States as part of President Nixon's ill-fated New Economic Policy. In economic theory it has always been maintained that price ceilings below free-market prices create shortages. And there wasn't an economist worth his salt who did not predict shortages as soon as Phase I of President Nixon's New Economic Policy was announced, including the chairman of the now-defunct Price Commission who had this to say: "The price system is far superior to any centralized model now available. The models are good theory, but not near at all to reality."[4] Of course, centralized economic models are *not* good theory. How can they be, if they don't work and are known not to work, and if their deficiencies had been exposed in numerous writings long before August 15, 1971? What's more, if the price system is now admittedly superior, was it not also superior in 1971? If so, why was it ever set aside?

The preceding quote brings up a very serious question: Why did the chairman of the Price Commission support price controls when in Washington and the free market when addressing petroleum industry executives in Houston? On the economist's inclination to give the advice he thinks his audience expects of him, see W.H. Hutt's book, *Politically Impossible . . . ?* (Institute of Economic Affairs, London, 1971).

The second moral of the story is related to the government's psychic fear of windfall profits. These are indispensable for the required stimulation of output that was discussed in connection with Figure 2-3. In fact, once the new equilibrium Point C has been attained, windfall profits will have vanished. They are of a transitory nature, and the government does not seem to realize this.

Transitional shortages can and do arise in free markets from increasing demand at a given supply as well as through declining supply at a given demand. Price controls will just as inevitably perpetuate transitional shortages in the one case as in the other. Perhaps this was the reason that a two-tier price structure was introduced in the domestic oil industry in the wake of the oil embargo.

To analyze the two-tier price structure, a brief look at the steel industry, where a similar but more obvious error has greatly contributed to the oil shortage, will be helpful. Having convinced the Cost-of-Living Council of the need for higher steel prices, the steel industry was granted price increases in sheet metals, but not in tubular goods. As any sophomore majoring in economics could have predicted, the steel industry cut back its tubular goods production and turned full steam to sheet metal, thereby aggravating the shortage of tubing, casing, line pipe, etc.—in the midst of an oil shortage! The relevance of the steel case is this: in the sector where price controls continued to be imposed, production effort was de-stimulated, only to be redirected into the higher-price sector.

This was and continues to be the precise effect of the two-tier price structure of crude oil. The system de-stimulates production of old oil, while stimulating the new oil production effort. The overall result is insufficient stimulation, since the weighted average price of U.S. crude oil remains below the market price and thus perpetuates the oil shortage. Nor does the two-tier price structure make much sense in regard to inflation, as will be seen in Chapter 3, where the inflationary impact of raising crude oil prices is discussed in detail.

As it stands, the price differential between new and old oil sets up strong pressures to cheat. In that area, too, the free market is not without ingenuity. To forestall the development of a black market in old oil, a new federal agency may be needed!

During the legislative debate on pricing policies of crude oil, the government, for once, listened to sound advice by at least exempting stripper wells from price controls. This was accomplished through a rider to the trans-Alaskan pipe line bill, but somewhere in the shuffle a stripper well was redefined as a well producing 20 bpd or less. If this new definition had gone undetected, over 30% of U.S. oil production would have been exempted as stripper oil. As it turned out, the White House noticed the "innovation" and insisted on the old definition of stripper wells (10 bpd or less). This reduced the exempt portion of U.S. oil production to about 12%.

It is now official U.S. policy to stimulate production of new and of ancient oil, and to de-emphasize the in-between (middle-aged?) oil.

iii. International Oil, the OPEC Cartel, and Windfall Profits

In 1970, some 22% of U.S. oil consumption came from foreign sources. By the first half of 1973, imports reached 35%, and one source has estimated total imports at 62% by 1985.[6] How the existence and market behavior of foreign suppliers affect price and total sales of oil in the United States depends on various factors. Domestic demand-and-supply conditions, including taxation levels, were discussed in section ii of this chapter. New elements are the cost of producing and shipping foreign-produced oil to the United States and the presence or absence of competition among international suppliers.

Foreign oil, if supplied competitively, is cheaper than U.S. oil. And at present U.S. consumption and import rates, there is a virtually unlimited short-term supply of foreign oil.

Assume, for example, that foreign oil could be delivered to the United States at a price of $2/bbl. in virtually any quantity.[7] In terms of a demand-and-supply model, this can be illustrated by adding a horizontal supply curve ($S_{foreign}$), Figure 2-4. Under competitive supply markets, the effect of opening the U.S. market to foreign supplies would be to reduce the price of U.S. oil from its indicated level of $3.10/bbl to, say, $2/bbl.[8] At that price, much of the U.S. oil-producing capacity would be wiped out. Domestic production rates would be given by Point E in Figure 2-4 (arbitrarily selected to show 8 MMbpd, and imports would be (again arbitrarily) 6 MMbpd, for a total U.S. consumption of 14 MMbpd.

This was the basis for the pre-embargo argument that the United States should open up its market to "cheap" foreign oil. This would cause the price of oil to be reduced instantly and permanently, so went the argument, and the U.S. consumer would reap the benefit. Put differently, the cheap-oil argument maintained that the U.S. consumer was forced to pay a higher price for oil than he would have had to pay in the prese ce of free markets. Given the asumptions on which the argument rests, it is irrefutable (even though refutations have appeared in print.)

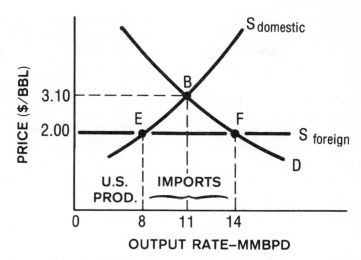

Figure 2-4. *The effect of competitive foreign supply markets on U.S. oil prices and output rates.*

The Real Costs of OPEC Oil

The "cheap" foreign oil argument was disproved with the formation of the Organization of Petroleum Exporting Countries (OPEC), which has become the most profitable cartel in the history of mankind. The obvious objective of forming a cartel is to raise prices or, at least, to keep them from being reduced. Accordingly, the first OPEC resolution called for maintenance of prices "steady and free from all unnecessary fluctuations."[9] Figure 2-5 shows effects on the U.S. of cartellization of foreign supply markets. It is a reproduction of Figure 2-4, with the addition of two possible foreign supply curves. Suppose, for example, the foreign cartel resorts to a unilateral increase of the U.S.-delivered price of oil from $2/bbl. to $2.50/bbl. (curve S′ foreign cartel). Such an upward shift of the foreign supply curve can be looked at two ways. On one hand it can be said that the overall price of U.S. oil is still lower than it would be in the absence of the foreign market; on the other hand, the price increase from $2 to $2.50 takes the form of monopolistic profits that accrue in part to foreign producers. It is extremely important, but generally overlooked, that the U.S. petroleum industry would not derive monopolistics long-term profits from a price increase imposed by foreign cartels. Short-term economic rents would, of course, accrue to the oil producers, but these rents would eventually vanish. That is, the oil industry would settle back on its supply curve, at the higher price and cost, where it would again be subject to a competitive rate of return. The competitive domestic market structure prevents the U.S. oil industry from shifting that curve up.

Foreign industries, and especially foreign government-controlled industries, are not subject to U.S. antitrust legislation; thus they can create a monopolistic market if the market itself permits it. This is precisely what OPEC has done, with the result that this organization obtains truly monopolistic profits, as opposed to the domestic U.S. oil industry, whose high-profit (high-rent) situation is transitory and, more importantly, acts as a natural, vital incentive for expansion.[10]

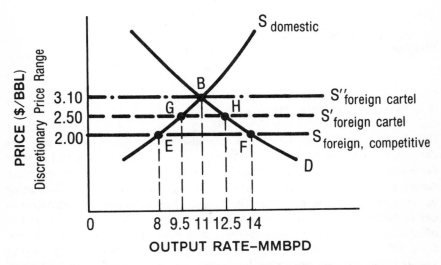

Figure 2-5. *The effects on the U.S. of cartellization of foreign supply markets.*

The hypothetical upper limit in the price-setting ability of foreign exporters is described by Point B, Figure 2-5. If the foreign cartel raised its prices to $3.10/bbl. (S″ foreign cartel), it would price itself out of the market.[11] This means the entire U.S. oil consumption of 11 MMbpd would come from U.S. sources. Thus, the foreign cartel's discretionary price range is defined by the competitive cost of producing and delivering oil to the United States at the bottom, and by the autonomous domestic price structure at the top. In terms of Figure 2-5, that discretionary price range falls between $2 and $3.10/bbl.

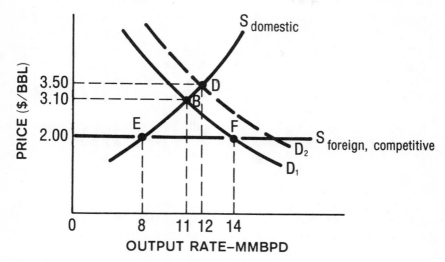

Figure 2-6. *Export cartels and increasing demand.*

The discretionary price range open to the foreign cartel widens over time due to increasing U.S. demands, as shown in Figure 2-6. Let Point B again reflect the purely domestic price structure and let $S_{foreign}$ reflect the competitive price floor. An increase in demand (D_2) that would raise the domestic price of oil to \$3.50/bbl. will also increase the discretionary price range. For any given price in that range, an increase in demand will increase oil imports. To the foreign cartel, the optimum pricing problem becomes one of weighing the benefits of a price increase vs. the loss in sales that will be induced by that price increase. Wherever that optimum price will be, one thing is clear: overall U.S. oil prices will be pulled up by the upward floating autonomous price structure that would prevail in the absence of foreign markets.[12] Unfortunately (from the U.S. point of view) or fortunately (for the foreign oil producers), increases in demand are not the only forces that tend to drive up autonomous U.S. oil prices: declining U.S. supplies will do the same thing. This case, implying a leftward shift of the supply curve, is not shown graphically. Of course, the term "declining U.S. supplies," in its economic meaning, is not the same thing as declining reserves; rather, it means harder-to-find, costlier oil.

The reason foreign oil prices have been pushed up so rapidly in the recent past is that the United States is faced with a simultaneous exposure to increasing demands and dwindling supplies. The limit to the price foreign producers can achieve in the United States is a function of the availability of oil substitutes in the production and consumption of energy.

One more point deserves mention. When a stripper well is shut down for a considerable period of time, it may never come back as a producer. That also holds true for other wells, such as those subject to natural or artificial water drives. Thus, an immediate reduction of import prices and the subsequent shutting down of U.S. oil wells will cause the loss of many wells to the U.S. economy. To give an idea of the order of magnitude involved, 12% of domestic oil production now comes from stripper wells.

A permanent destruction of U.S. oil wells through temporary cutting of prices would push the U.S. supply curve to the left as shown in Figure 2-7 $S'_{domestic}$. If the price were subsequently raised back to the level that prevailed prior to the price cut, less domestic oil would be produced than before. If the same domestic production rate were desired, a greater outlay of money would be required to achieve it. This, in itself, would put a foreign cartel in a position to raise its prices (and the U.S. price of oil) beyond the current domestic equilibrium price by first cutting and subsequently raising the world price of oil. Thus, "cheap" oil could, in fact, become very expensive.

In terms of Figure 2-7, let the foreign cartel reduce its price to the competitive level of \$2/bbl.[13] Six months or so later, the deterioration of shutdown U.S. wells would have shifted the domestic supply curve to the left ($S'_{domestic}$). The resulting domestic equilibrium price (no imports) would have risen to \$3.50, Point I, and the cartel can now charge a higher price than before. So much for the cheap-oil argument. As it turned out, U.S. fears that OPEC would resort to predatory price cutting were unfounded. The response of U.S. policy planners to rising world oil prices was so weak and so confused that OPEC's positon was strengthened, not weakened, after the embargo. Thus, OPEC never saw the need to engage in costly price wars.

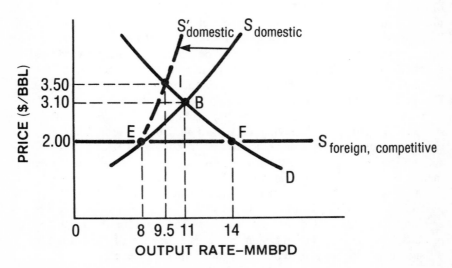

Figure 2-7. The permanent destruction of stripper wells.

The Dangers of Two-Tier Pricing

While windfall profits will eventually vanish, their life cycle is closely tied to the production cycle. Returning to our earlier analogy, if all but one hamburger stand burned down in a given city, the lone survivor would reap windfall profits for about a week. The demand-induced price increase would soon attract new operators who would hurriedly set up business, thereby destroying the survivor's temporary monopolistic profits.

Oil has a long production cycle, about five to ten years. Seismic work, exploratory drilling, field development, road building, pipe laying, regulatory hearings, environmental impact studies, construction of refineries and tankers, plus other factors tend to prolong the production cycle. Windfall profits can, therefore, last for a long time, so why not eliminate them through a two-tier domestic-foreign price structure or a special tax on "excess profits"? To resolve this question, it is absolutely necessary that the *oil* perspective be abandoned and an *energy* perspective be adopted instead. To simplify the discussion, it will be assumed that the only long-term remedy to the current energy shortage is nuclear fusion, a controlled hydrogen bomb. The theoretical feasibility of this process is well explored, but resolving the technological problems will require huge capital funds. Where are those funds going to come from? From the windfall profits! That is the very function of these profits: under a free pricing system, the greater the foreign pressure on landed oil prices, the greater these windfall profits and the greater the incentive to develop nuclear alternatives. This, in itself, would act as a constraint on foreign cartels providing that the U.S. policy planners adopt an expansive energy perspective, allowing oil companies to develop other sources of energy. Unfortunately, the legislature is pushing for vertical and horizontal divestiture of the U.S. oil industry, the very opposite of what is presently needed.

In his 1974 State of the Union Address, President Nixon announced that the federal government planned to spend $10 billion over the next five years on energy research and development, compared to a $200 billion investment program of private industry over this same period. As will be seen in the following calculation of windfall profits, the free market is capable of generating something on the order of $100 billion in economic rent over the next five years if—and only if—these funds are not legislated away by price controls or rollbacks, or by taxation.

The discussion of windfall profits has so far neglected quantitative aspects of the question. To remedy this situation, a case simulation is required. Such a simulation was run in early 1974, and the results are given here. However, it must be noted that certain more or less reasonable assumptions have been made for the sake of this simulation. These assumptions may or may not hold (in fact, as subsequent events showed, some of them did not). This does not really matter, for the simulation was not intended to be a projection of things to come. On the contrary, its purpose was to point out the questions that would eventually have to be answered and the direction that must be followed to develop a feel for the actual quantities involved.

The assumptions used in the case simulation were:

1. The foreign cartel price of oil was held at $10/bbl. for several years (other cases for different cartel prices must be run), and the overall cost of production for all OPEC countries was assumed to be $1.50/bbl.

2. For the two-tier case, an average U.S. price of domestic crude oil was assumed to be $4/bbl.

3. The life cycle of oil production is such that the average production cost will remain at $4/bbl. for at least five years. (This is one assumption that must be modified with a three to five year lag mechanism in an extended model.)

4. The U.S. supply of oil continues to decline by about one million bpd per year.

5. Approximate production volumes and prices are taken from the data for the first half of 1973, the pre-embargo data: approximately 17 million bpd consumed, with 11 million of U.S. and 6 million of foreign origin, at $4.00/bbl.

6. A downward adjustment in demand by one million bpd due to the sudden 1973 price increase in crude is assumed to take place over a two-year interval. (This assumption also needs modification.)

Results of calculations based on these assummptions are shown in Tables 2-2 and 2-3. For the first year, it is immediately apparent that the consumer pays less—a lot less—under the two-tier price structure.

On the other hand, the foreign cartel gets substantially higher profits under the two-tier price structure ($12.4 billion vs. $6.8 billion). Why? Because the low domestic price of oil ($4/bbl.) discourages U.S. production and the cartel picks up something like 2 MMbpd that would have been produced domestically at a U.S. price of $10/bbl. Similarly, the U.S. dollar drain (shown as OPEC revenue in Tables 2-2 and 2-3) and thus the downward pressure on the U.S. dollar are much higher in the two-tier price case. The calculated $14.6 billion drain in year 1 compares to an overall balance-of-payments deficit (liquidity balance) of $13.6 billion in 1972.

Finally, and most importantly, no funds are being generated for nuclear or other energy development in the two-tier case, whereas the free-price case generates some $28 billion the first year.

Even more surprising are the long-term trends revealed in the tables. First, the total cost of oil to the U.S. consumer rises steadily in the two-tier case because domestic production is held back, and declining U.S. producing capacity calls for greater and greater amounts of high-priced foreign oil.

In the one-price case, at the stipulated average price of $10/bbl., the total cost to the consumer declines at first. This reflects the long-term adjustment in consumption patterns discussed previously. Once this adjustment is

The Energy Crisis   35

Table 2-2
Theoretical Effects of One Price of Oil Over a Five-Year Period

| | Consumption (million bpd) | | | Cost to U.S. Consumers (billions of $) | | | | | | |
| | | | | | Revenue in U.S. | | | OPEC Revenue | | |
	Total	U.S. Oil	OPEC Oil	Total	Total	Prod'n Cost	Energy Development Fund	Total	Prod'n Cost	Cartel Profit
Year 1	15.0	12.8	2.2	$ 54.7	$ 46.7	$18.7	$ 28.0	$ 8.0	$1.2	$ 6.8
2	14.0	12.0	2.0	51.1	43.8	17.5	26.3	7.3	1.1	6.2
3	14.0	11.0	3.0	51.1	40.1	16.0	24.1	11.0	1.6	9.4
4	14.0	10.0	4.0	51.1	36.5	14.6	21.9	14.6	2.2	12.4
5	14.0	9.0	5.0	51.1	32.9	13.2	19.7	18.2	2.7	15.5
Total				259.1	200.0	80.0	120.0	59.1	8.8	50.3

Table 2-3
Theoretical Effects of Two-Tier Pricing of Oil Over a Five-Year Period

| | Consumption (million bpd) | | | Cost to U.S. Consumers (billions of $) | | | | | | |
| | | | | | Revenue in U.S. | | | OPEC Revenue | | |
	Total	U.S. Oil	OPEC Oil	Total	Total	Prod'n Cost	Energy Development Fund	Total	Prod'n Cost	Cartel Profit
Year 1	15.0	11.0	4.0	30.7	16.1	16.1	0	14.6	2.2	12.4
2	14.0	10.0	4.0	29.2	14.6	14.6	0	14.6	2.2	12.4
3	14.0	9.0	5.0	31.4	13.1	13.1	0	18.3	2.7	15.6
4	14.0	8.0	6.0	33.6	11.7	11.7	0	21.9	3.3	18.6
5	14.0	7.0	7.0	35.8	10.2	10.2	0	25.6	3.8	21.8
Total				160.7	65.7	65.7	0	95.0	14.2	80.8

made—when the consumers operate on the higher-elasticity long-term demand curve—total consumption stabilizes. At the fixed price of $10/bbl., this means stable but high total costs to the U.S. consumer.

In both cases, two-tier pricing and one price, foreign cartel profits rise with time. Moreover, they rise about the same amount, i.e., by some $9 billion over the first five years. But since the two-tier case provides the cartel with a considerably higher initial profit, the total five-year profit is also much higher in that case: $80.8 billion, compared to $50.3 billion in the one-price case. A similar pattern develops for the U.S. dollar drain as it relates to payments balances.

Again, no funds are provided from within the energy sector for nuclear or other energy research if the two-tier price system is maintained under the assumed conditions. The total five-year funds generated under the one-price system amount to $120 billion under the assumed circumstances. But, and this is a frightening reality, these annual fund flows decline rapidly, in line with the decline of U.S. oil production. Thus the U.S. government has a solemn obligation not only to introduce the one-price system but to do so *before it is too late.* For advice on what to do after it is too late, the reader is urged to consult with the governments of Japan or Western Europe.

As was mentioned in the discussion of Figure 2-7, foreign energy cartels are capable of wiping out a substantial portion of the U.S. producing capacity, certainly all marginal wells, perhaps all pumping wells by a judicious use of price-war tactics. What's even more frightening is that such tactics can also wipe out any and all previous investments in nuclear or other energy research.

For example, if oil prices were to be cut to $2/bbl. after the fifth year of operation of the one-price system, $120 billion worth of research would be in jeopardy. The U.S. government must be aware of this, but it must also keep in mind that a subsequent round of monopolistic price increases is likely to start as soon as the research lead has been annihilated. Therefore, the U.S. government must be firmly resolved to continue its nuclear research effort, no matter what, once it is underway. In a price war, you can close a filling station and walk away from it, even repeatedly. You cannot do that to $120 billion worth of nuclear research.

The destructive capability of foreign cartels compels the United States irreversibly to a policy of searching for nuclear alternatives. The greater the cartel price of oil, the more intensive this research effort will have to be, and the sooner a long-run solution will be found. Moreover, the United States will likely share its technological expertise with other nations, and that, in the long run, could end the oil cartel.[14] Certainly cartel members know this, and it is to be hoped that they recognize the trade-off they are facing between short-lived superwealth and sustained prosperity.

While it is true that unreasonably high oil prices will stimulate U.S. and European research for energy alternatives, all but the nuclear fusion solution are stop-gap measures. They will do no more than to delay the day of total nuclear energy dependence, and the nuclear fusion process is decades away. Whatever interim alternatives are used (oil shale, tar sands, synthetic oil or gas from coal, coal itself, nuclear fission under the conventional U-235 reactor or the breeder reactor), it is absolutely impossible for the United States to be totally self-sufficient in energy by 1980, as President Nixon hoped it would be in his announcement of Project Independence. Chances are the President knew this and merely meant to put pressure on OPEC to abandon their monopolistic policies. Such an approach is dangerous, since it may be interpreted as a bluff and thus a sign of weakness.

In the cartel simulation, an arbitrary assumption was made concerning a cartel price of $10/bbl. Is there not a ceiling on the price that can be charged by the foreign cartel? Of course, such a ceiling exists. It is determined by the cost of producing substitute forms of energy at required output rates. This cost is itself a function of the U.S. attitude toward many things, but especially toward pollution. The clean air vs. cheap energy dilemma, one of the greatest stalemates of our time, has been discussed ad nauseam elsewhere and will not be taken up here for that reason.

No one seems to know what the ceiling price really is, except that it is always higher than anticipated. Nor can anyone be blamed for this knowledge gap. New techniques and processes need to be developed to meet the crisis, and some of these techniques are highly sophisticated. The energy companies are moving in virgin territory and cost estimates will firm up with experience. Currently, however, uncertainty prevails.

OPEC's Economic Might

In the discussion of oil-producing cartels it was pointed out that their ability to implement noncompetitive market strategies depends on their economic power. Several factors have a bearing on this power:

1. The importance of foreign cartel-held reserves, both geographically and by volume;
2. The market share of the cartel in the United States;
3. The ability of the cartel to survive if all oil exports are cut off or threatened to be cut off;
4. The ability of the United States to survive if all imports from the cartel countries are cut off or threatened to be cut off;
5. The determination of cartel members not to "cheat" each other, i.e., not to make secret concessions to purchasers.

No matter how important foreign cartel-controlled oil reserves are, if the cartel's U.S. market share is insignificant, it has relatively little influence on domestic markets. On the other hand, a large foreign cartel commanding a substantial share of the U.S. market holds considerable market power, especially if geographic coverage of the cartel is complete (if U.S. producers have no place else to go). Recently, the trend has been toward larger and larger market shares held by OPEC in the United States. This implies greater and greater economic power and a further and more forceful implementation of noncompetitive market strategies. Also, this development is paralleled by increasing exports of OPEC oil to Western Europe and Japan, and by increasing revenues to OPEC from oil-derived non-oil investments which have increased OPEC's viability in the event of, or under the threat of, U.S. economic retaliation. Today, the OPEC countries' economic independence of the United States is complete.

As was pointed out earlier, OPEC is in no immediate danger of breaking apart of its own weight. Strong pressures that *tend* to break up a cartel always exist, but strong pressures also exist to keep it together. Which of these is the stronger force is an open question. Be it from an observation of U.S. markets, or as a result of the teachings in U.S. universities, there is a presumption in the United States that the disruptive power will ultimately prevail. With the help of antitrust laws this may be true in the United States, but it is simply not true generally. Cartels have survived for centuries in Europe, and one of the less successful battles in the European Theater was the U.S. attempt to break German cartels after World War II.

United, cartels can and will survive; divided, they will perish. Whatever one may say about OPEC, its founding members were aware of this fundamental problem. To quote again from their first resolution: ". . . no . . . member shall accept any offer of a beneficial treatment, whether in the form of an increase in exports or any improvement in prices. . . . "[15]

Western Myopia

In fairness to the OPEC countries, they are not the only cartel in the world energy market. The oil-producing countries have been contending with at least two cartels, one of of which survives to this day. The surviving cartel is actually a conglomerate of independent cartels, each endowed with absolute economic power within certain well-defined boundaries.

These cartels are the governments of the oil-consuming countries, in particular the European governments. Through exorbitant levels of revenue-producing taxation they have driven up prices of oil products, particularly of gasoline, and thereby confiscated the lion's share of the value of finished oil products. This is forcefully evidenced in Table 2-4.

As the table shows, the consuming countries' take of the oil was *six times as high* as that of the producing countries and 7½ times as high as that of the oil companies. And the consuming countries did not, as did the oil companies, put out any effort in the productive process of providing energy for their people.

Table 2-4
The Price Component of One Barrel of Oil

(Based on the 1967 average price of one barrel of oil sold in Western Europe)

	Absolute Value	Percent of Total
Cost of production	$ 0.285	2.7
Cost of refining	0.350	3.3
Tanker freight	0.680	6.3
Storage, handling, distribution and dealer's margin	2.790	26.0
Oil company's net profits..................................	0.681	6.3
Indirect and turnover oil taxes in consuming countries	**5.100**	**47.5**
Revenue of producing countries	0.853	7.9
Totals	$10.739	100.0

Source: Kayal, A.D., *The Control of Oil: East-West Rivalry in the Persian Gulf,* a dissertation, University of Colorado, 1972, p. 266. Basis was OPEC data.

In retrospect, one cannot help but wonder why this message was not received in Europe. Whether this was the result of OPEC's lack of experience in Western-style public relations or simply because Europe was not prepared to receive the message is a moot question. This type of exploitation had been going on for years until the OPEC countries resorted to unilateral actions of their own. If the European countries had shown a little understanding and foresight at the time, they might have agreed to let OPEC countries have at least a revenue equal to theirs by simply reducing tax rates in Europe and letting producing countries pick up the difference. This would not have driven up gasoline prices in Europe; such an adjustment would have peacefully recognized the surging economic power of the oil-producing countries. Instead, European intransigence forced a policy of economic warfare on the OPEC countries.

The second cartel, no longer alive, was established by the U.S. government through its import quota system. To recognize its perniciousness, this system must be viewed from the perspective of oil-producing countries. From their point of view, oil was being shipped to the United States at artificially reduced prices. These prices were held down or, rather, U.S. oil prices were maintained above world oil prices, through import quotas.

The world's foremost free-market country thus denied the oil producing countries free access to U.S. markets. To add insult to injury, the quota system openly permitted oil importers (quota holders) to confiscate at least a portion of the differential between United States and world prices by establishing a secondary market in import quotas. And while a controversy raged in America over whether this system would help or hurt the U.S. consumer, no one asked how it affected the Arab consumer. Indeed, the term "consumer" applied to Arabs is still foreign to American thinking today.

Necessary Perspectives

Neither Europe nor the United States has gone to a great deal of trouble to understand the thinking of their trading partners, and this is still a sorely neglected area that needs more attention. Even the Cabinet Task Force Report on *The Oil Import Question* never so much as hinted at the problem. Indeed, the American academic and business communities are more, not less, insensitive to OPEC thinking than Europe is. This is one area that could stand a great deal of improvement.

It may be too late now, but years ago the oil industry should have set up an international research center where scholarly observers not directly linked to the oil industry could monitor the thinking of the foreign petroleum elite. Such a center would have communicated long ago that leaders of developing countries, with the support of academic leaders throughout the world, hold nationalization to be legal, as does the U.S. Department of State. Whether it *is* legal under international law is totally irrelevant. It is perceived to be legal and that perception is supported by an abundant literature. The point is that foreign leaders as well as their American counterparts will make decisions not on the basis of what *is* right, but on what they *perceive* to be right. Unfortunately, American and foreign perceptions differ, and they drift more and more apart.

One very practical insight might have been gained by such a research center. The man in the street, as well as the U.S. petroleum industry, seemed to believe that the oil embargo was temporary, i.e., that all would be as before, after the embargo had been lifted.

The fact is that the world has changed permanently. First, prices will never come down to pre-embargo levels. And second, after the initial political embargo, there followed a partial economic embargo. Indeed, one of the most difficult decisions facing the governments of the oil-producing countries is to determine the optimum output rates of oil, regardless of the political situation. Whatever ouput rate this may be, and it will vary according to the different countries' internal development potentials, it will certainly be less than the volumes that Europe, Japan and the United States will claim.

The policy of reduced output rates is predicated on more than just the belief that rising prices make oil in the ground appreciate faster than money in the bank. The problem of optimum production cannot be solved here in one or two sentences, nor can the case for restrained output be rejected. It is a mistake flatly to reject the notion of reduced production rates by calling it "a parade of mistakes."[16]

A question that the oil industry should have asked but failed to is: If *we* were the economic advisors to an OPEC government, what production rates would we propose as being in their long-term best interest? Such a simulation would have had the virtue of sensitizing the American oil industry to the problems of their foreign partners: by forcing the Americans to focus on this problem from the partner's point of view they would have had a broader understanding of the issues at hand. Certainly, they would have known that the political embargo would be followed by a partial economic embargo.

Similarly, and more importantly, OPEC members need to come to grips with American and European alternatives. They need to ask questions concerning the political mood and determinations of the oil consumers and their legislatures as well as the technical feasibility of alternative energy possibilities. "If I were President of the United States, what would I do and what could I do in response to a further tightening of OPEC oil supplies," is one game that OPEC members ought to play.

A detailed discussion of the world supply and demand situation is beyond the scope of this chapter. Suffice it to say that the OPEC cartel would indeed be seriously weakened if oil became plentiful again. But such a development is doubtful. It would take something on the order of another Middle East two times over to inpart a temporary malaise among OPEC members. Only Indonesia holds out hope of being a future Middle East, and Indonesia, of course, is an OPEC member. As for the North Sea, its oil will always be expensive. Venezuela's production has peaked out, and the prospects of developing another Middle East in South America are dim at present.[17]

In summary, the energy outlook is grim, not only today but for decades to come. One aspect of the energy crisis is a general decline in wealth throughout the Western World which will be felt in all areas: food, clothing, travel, education, leisure. Policy mistakes have contributed to, and continue to contribute to this dilemma. While there is hope that a low-cost substitute for fossil fuels will eventually be found, there should be no illusion concerning the severity of the problem until that time. Instead of allowing the various oil interests, exporters vs. importers, to polarize in opposing camps, an honest attempt at two-way international co-operation will be beneficial for *both* sides. In the absence of a spirit of co-operation, the current economic conflict may eventually lead to armed conflict, and that would be the most serious of all mistakes.

iv. **Oil & the Decline in U.S. Living Standards***

Today's number one crisis, energy, is a veil. Everybody talks about it because everybody can understand it. Yet, pressing as it is, it conceals a much more important, largely ignored long-term problem—the declining wealth of our nation.

---

*This section has been co-authored by Dr. X. Boisselier, Director of the Institut d'Administration des Entreprises, University of Nice, Nice, France.

The term "wealth," as used in this context, denotes the living standard a people can attain: the total number of goods and services a nation produces in a given year and thus makes available to its citizens. Leaving out the problem of capital-goods production, either as replacements for worn-out capital goods or in the form of new net investment, the Gross National Product (GNP) is a reasonable measure of a nation's wealth.[18] To attain a given GNP, a nation's economy must have the ability to produce the corresponding number of goods and services, and it must also have the purchasing power and ability to take its total production off the market. In short, the nation's aggregate supply (AS) and aggregate demand (AD) must be in equilibrium. Variations in either AS or AD bring about disturbances in the economy.

There is an essential difference between AD and AS problems. The former are easily recognized and dealt with through appropriate monetary and fiscal policies. Aggregate demand problems have been thoroughly explored, analyzed and discussed in the economic literature. For this reason, they will be passed over in this discussion. Any good book on macroeconomics or any worthy business magazine or newspaper (*Fortune, Business Week, Baron's, The Wall Street Journal,* to name a few) deals, oftentimes implicitly and sometimes unknowingly, with the nation's aggregate demand. This does not hold true for the nation's aggregate supply, which is being discriminated against. Its existence is commonly ignored. What's more, AS-induced economic disturbances are not easily remedied.

Aggregate supply refers to the nation's ability and willingness to produce. If production is deliberately withheld, the resulting contrived scarcity of goods and services causes the GNP to decline while prices rise. Willful witholding of production may originate in one of several sectors of the economy:

Witholding of goods and services from the retail or intermediate goods markets. This is the case of the well-known monopoly or cartel.

Withholding of goods and services from input markets, which can take one of two forms:

1. A buyer dominating the market with power to depress input prices (a monopsony). Witness the monopsonistic hold of Alcoa on bauxite ore throughout the world, prior to their breakup by antitrust litigation.

2. A union (in the case of withheld services, especially labor services). The term "union" is used here in its broadest sense to include all restrictions on entry, whether a union is formally created or not. Under such a definition, the American Medical Association would be a union, since it determines admission requirements for medical students. Similary, the American Bar Association would be a union.

Contrived scarcities may arise when there exists a negation of the nation's willingness to produce. Technically, the nation's aggregate supply is deliberately held down. The upshot of such an artificial reduction in AS is a stagflation: the GNP declines while prices rise. For brevity, the following is a simplified explanation of stagflation. Willful withholding of production causes the output rate of goods and services in a nation (the GNP) to decline. Given the nation's money supply, goods are now scarce in relation to money. As a result, they demand a monetary premium: prices will rise. In more popular terms, existing dollars will "chase" fewer goods, causing goods to attract more dollars. Thus, a stagflation results from any form of willfully withholding output, and with it a decline in living standards. Put that way, the antisocial behavior of all non-free-market power centers (monopolies, monopsonies, unions) clearly emerges. There ought to be legislation against this kind of behavior, and there is.

Monopolies and monopsonies can be and have been broken up through antitrust litigation. No equivalent legislation exists to protect the American public from unions.[19] Thus, as unions become more and more "successful" in enforcing their wage demands (as well as in their implicit wage demands such as fringe benefits, cost-of-living escalators, etc.), the contrived scarcity of goods causes the GNP to decline.

Here, then, is the first drain on the nation's wealth. It is a drain that has gone fairly unnoticed, one that assumed really damaging dimensions in the late 1960's. But this is not the only reason for the nation's declining wealth.

In additon to the deliberate reduction of a nation's aggregate supply, its ability to produce may also be impaired by the depletion, or partial depletion, of its natural resources—any resources, including oil. Energy is one U.S. resource that is currently being rapidly depleted. As contrasted with the contrived scarcities previously discussed, this is a case of natural scarcity. Mistaken economic policies may hasten the depletion of natural resources, but that is all these policies are capable of doing. The day of reckoning will eventually come, with or without economic policies.

The problem is that nothing can be done to stem the resource-induced reduction in the nation's AS, or almost nothing. Insofar as energy is concerned, the one hope to get out of the dilemma is the development of an equally cheap substitute source of energy, ultimately nuclear fusion. This and the implicit need for research funds, have been discussed in section *iii*.

Unfortunately, public opinion still considers the energy shortage a temporary phenomenon, and most writers continue to express this opinion. In one editorial appearing in an oil trade paper, the author discussed "ideas for overcoming the energy crisis for this country."[20] His two prime requisites for capturing untold billions of barrels of domestic crude oil and trillions of cubic feet of domestic natural gas are "time and incentive." The author never fully addresses himself to the aggregate supply question, probably because he fails to recognize that the nation's wealth is at stake, and permanently.

Certainly, the preceding remarks call for elaboration.[21] To explain the problem, the following discussion of the nation's declining wealth will be real rather than monetary terms. Furthermore, the discussion will be in to oil for simplicity. The argument can easily be expanded to include natural gas, or all forms of energy, or any other natural resource. The cost of finding and producing oil will be expressed in terms of productive resources that must be allocated to the task and that are, as a result, no longer available to produce other goods.

The reduction of oil reserves affects the nation's total output or GNP in the same manner as the partial destruction of its capital equipment or (Heaven forbid) the loss of a part of the labor force by a natural disaster. Oil, very much like capital equipment and labor, is a vital input in the production of a nation's GNP. Everything else remaining the same, a sudden decline in oil reserves can only result in a reduced GNP. There is no other way.

Superficially, it would appear that the United States has three alternatives in meeting the current oil shortage (or that it can combine the three):

1. Let oil prices rise.
2. Grant tax incentives to the U.S. oil industry.
3. Increase oil imports.

The case for higher oil prices was the subject of sections *ii* and *iii,* and it will be taken up again in Chapter 9. Suffice it here to say that it is the only alternative acting on both the demand for and supply of oil. Being the most efficient allocation mechanism, the price alternative minimizes the negative wealth impact of declining oil resources, but it does not totally eliminate this impact.

In regard to the tax-incentive case, it will be assumed that there is no waste in the administration of the system and that the system is geared to reward for oil discovery, not simply for drilling. A type of expanded depletion allowance would do just that. Such a measure could stimulate exploration and production activities to the extent that the U.S. could be producing all of its projected oil needs from its own soil. Certainly, given a sufficiently high national priority, that is theoretically possible. But it is also costly.

Since oil is increasingly more difficult to find, and oil pools are becoming both deeper and smaller, or are located offshore or in the Arctic, drilling and transporting it to markets would become more and more expensive and the success ratio of commercial strikes would decline rapidly. Leaving out the lag problem, it is probably safe to say that a sustained doubling of U.S. oil production by 1985 would require something like four to six times the exploration effort of recent years.

To determine the locations and to drill the needed 30,000 exploration wells year after year requires a substantial reallocation of men and machinery away from some other goods (say, refrigerators) and into the oil industry. Thus, the cost of filling the oil gap domestically is the non-production of so many refrigerators; refrigerators that once were, but are no longer, available in the economy. In short, the tax-incentive route will stimulate oil production, but it inevitably brings with it a decline in the availability of other consumer goods, or the living standard of the people.

The notion that the stimulated increase in oil production makes up for or even more than makes up for the decline in consumer goods (say, through an increase in energy-related productivity), is untenable. This is an illusion held by countless people and one that has been carefully nurtured by the media. The fallacy of this notion is illustrated by the following example in which oil consumption, and with it the nation's productivity, is assumed to remain constant between now and 1985, at a level of 16-17 million bpd. Also, a realignment of U.S. priorities is stipulated such that the entire U.S. oil consumption is produced domestically.

This also would require a more intensive exploration program, perhaps a doubling of the drilling rates of recent years. Again, even though the United States would now only be trying to perpetuate the status quo, it would still have to shift resources away from consumer goods and into oil. And again, there is no escape from the fact that the nation's GNP would decline as a result.

But what about imports? Will that not solve the problem? Cannot the United States maintain its domestic allocation of resources by importing the required oil? That would enable the United States to have both refrigerators and oil, or would it? Unfortunately, it would not.

If, under the stipulated circumstances, the United States only maintains its current exploratory efforts while perpetuating its current oil consumption rate of 16 to 17 million bpd (zero-increase in productivity), its failure to offset dwindling oil reserves with more intensive exploratory efforts will widen the domestic oil gap, say, from the pre-embargo 35% rate to something like 50%. Imports, in other words, will rise from 6 million bpd to about 8.5 million bpd or more. Since no resources had to be allocated to additional oil exploration and drilling, the United States would be producing the same amount of consumer goods as before. But—and here lies the problem—not all of these would be available for U.S. consumption.

Imports are ultimately paid for by exports. If, therefore, the United States had to pay for oil imports of 8.5 million bpd by exporting refrigerators, it would take some 23 million refrigerators annually to pay for these imports, assuming the now obsolete pre-embargo prices of $400 per refrigerator and $3 per barrel of imported oil. Actually, the U.S. economy produces only 6.3 million refrigerators annually, so that its oil imports would have to be financed by exporting additional goods.

Again, using the pre-embargo price of $3 per barrel of oil, financing the importation of 8.5 million bpd from abroad would involve the following production effort, or its equivalent: the entire U.S. output of dryers, washers, dishwashers, freezers, ranges (both electric and gas), refrigerators, black-and-white TV sets, plus 62% of the U.S. output of color TV sets, based on 1972 data as listed in *U.S. Statistical Abstracts,* 1973 edition, pp. 734 and 735. Such is the wealth of the petroleum-producing countries.

The United States, as a result of its oil imports, would find itself producing everything as before, but it would no longer be consuming the items listed above, or their equivalent. Certainly the U.S. material well-being or living standard would suffer. There is no escape: A decline in U.S. oil resources is tantamount to a decline in its national wealth. Price increases of cartel-held foreign oil aggravate the situation because it now takes more refrigerators or other U.S. goods to import a given volume of oil.[22]

The most disturbing aspect of this second attack on the nation's wealth, through a depletion of its natural resources, is the fact that the permanence of this state of affairs apparently has not been recognized. Fate has decreed that oil should be the first such natural shortage. Others will follow. Worse, while there is hope of replacing the oil-induced energy shortage through nuclear energy in a generation or so, there is at present little hope of finding suitable substitutes for other resources such as copper, lead, zinc, to name a few.[23]

It seems reasonable to assume that the oil cartel will be emulated by cartels in other resources as those resources become increasingly scarce. After all, the formation of cartels and the use of export embargoes as a policy tool is not uncommon. The U.S. embargo on various agricultural commodities, imposed to combat domestic inflation, is an example of unilateral government action in this area.

A military response to an economic squeeze seems irresponsible and downright foolish unless the United States is prepared to seize and hold permanently all resource bases in the world. Some of the countries that would have to be invaded, and their percentages of world trade in scarce resources, are:

- Copper—Chile, Zambia, Zaire and Canada (74%).
- Lead—Australia, Mexico, Peru and Canada (90%).
- Tin—Malaysia, Bolivia and Thailand (78%).
- Zinc—Canada and Mexico (62%).
- Aluminum—Jamaica and Surinam (70%).

(Source: *International Financial Statistics,* IMF, September 1973, pp. 28-1)

In addition to the basic raw materials listed above, the United States is currently importing more than 50% of its domestic consumption of chromium, manganese, and nickel. By 1985, iron and tungsten are expected to follow suit, with potassium joining by the year 2000.[24]

In regard to oil, many influential writers have publicly advocated use of force for the protection of "inherent rights." The year was 1973; the country involved, Libya; the shortage, oil. In 1936 Hitler invoked these same rights vis-à-vis Eastern Europe for the alleged shortage of Lebensraum. Will humanity ever learn from history?

Apart from the moral problems involved, it is obviously impossible to seize, let alone hold, all countries possessing scarce resources. It cannot be done using conventional weapons, and it cannot be risked with nuclear weapons. And even if the United States embarked on the course of military intervention, it would only forestall the day of final reckoning, for resources still are going to be depleted. Too, such a forceful rearranging of the world-wide flow of resources, with the United States being the primary recipient (probably at artificially depressed prices, if military occupation does indeed take place), would open up the United States to charges of world-wide exploitation (the technical term used by the other camp is "neocolonialism").

This is not to suggest that U.S. oil companies have been operating under the military umbrella of their government. On the contrary, the record clearly indicates that the U.S. Department of State has by and large pursued a hands-off policy vis-à-vis foreign investments of private U.S. capital.

Still, the fact remains that 6% of the world's people, living in the United States, consume 30% of the world's energy. Whether this is a sound basis for the doctrine of capitalistic explotitation is questionable, very questionable. But the figures do imply that energy is underpriced in the United States. If this is the case, energy is indeed being squandered here, and other natural resources may be, too. Most people in the world, and some in the United States, think so.

Here, then, is the second drain on the nation's wealth. The first, it will be remembered, was the contrived scarcity of goods and services caused by monopolies, monopsonies, and especially unions. The second is the result of the nation's depletion of natural resources. The first kind is difficult to combat, mostly for political reasons that time does not permit to discuss in detail. The second kind can be overcome only by the development of low-cost substitutes. In the case of energy, this may be a slow process, measured in decades rather than years.

The issuing of fake money for domestic as well as international use can be used, and has been used, by the U.S. government (and others) to buy time, perhaps in hopes that the problem would go away. Monetary and fiscal policies have held the U.S. economy at or near the full-employment level, but as contrived and natural scarcities in this country are becoming more and more serious, more and more stimulants are required. All Americans now are paying the price for these policies, by way of rising price levels. What's more, since there is no sign of an easing of shortages, general price increases are bound to continue, more and more rapidly.

The oil embargo aggravated this situation, and it forced the U.S. government to inject the economy with more stimulants, with the result of an ever-increasing inflation rate. Even more ominously, OPEC-like organizations have been formed or are now being formed in aluminum, copper, tin and uranium. Surely others are being contemplated by people who are in a position to implement them.

Thus, the future looms menacingly on the horizon. In regard to our inability to come to grips with the wealth decline induced by contrived scarcities, we have no one to blame but ourselves. That part of the wealth decline resulting from diminishing resources is inescapable, though it is accelerated by mistaken government policies.

One easy way out, not in solving the problem, but in finding a scapegoat, would be for the U.S. government to blame the whole thing on the Arabs, both the contrived and the natural scarcities. Whatever economic policy is followed thereafter, the government could plead "not guilty" to the consequences: a continuation of easy-money policies would greatly accelerate an already intolerable rate of inflation; tight monetary and fiscal policies would plunge this country into a deep depression. In either event, a convenient scapegoat is at hand.

References

1. Othman, T.M., Harris Trust and Savings Bank, Chicago, "Corporate Foreign Currency Hedging Strategy . . . The How and Why," in *The Futures Market in Foreign Currencies,* International Monetary Market, p. 33.
2. Corm, G., "Arab Capital Funds and Monetary Speculation," *Arab Oil and Gas,* August 16, 1973, pp. 24-30.
3. The Chase Manhattan Bank, *Outlook for Energy in the United States,* June 1972, p. 44.
4. *The Oil Daily,* Nov. 14, 1973, p. 8.
5. *The Oil Daily,* Oct. 24, 1973, p. 2.
6. Wright, M.A., *Statement Before the Governor's Conference on the Big Cypress Swamp;* May 17, 1971.
7. Production cost in North Africa and the Middle East has been estimated to be 10-20 cents per barrel. Shipping cost from the Mediterranean to the U.S. coast has ranged from $.53 to $2.77 per barrel. The latter figure reflects a temporary tanker crunch.
8. Prices and quantities in the figures are hypothetical and are not meant to reflect or suggest actual or potential market conditions.
9. As quoted by J.E. Hartshorn: *Politics and World Oil Economics,* Revised edition, Frederick A. Praeger, Publishers, New York (1967), p. 27.
10. In 1973, 1,152 private antitrust suits had been filed in federal courts. In addition, the Justice Department filed 62 antitrust cases, compared to some 30 filed by the FTC. As can be seen from these statistics, antitrust legislation is not a dead issue in the U.S. *The Wall Street Journal;* Nov. 29, 1973; p. 1.
11. To prevent misinterpretation it should be emphasized once again that the prices shown in the figures are hypothetical. Certainly, the short-run price ceiling is sub-

stantially higher than the indicated $3.10/bbl. The precise level of this ceiling is governed, among other things, by the cost of other energy alternatives. This has been correctly analyzed by the OPEC countries and has caused them to conduct a study on price-supply curves of alternative sources of energy. *Middle East Economic Survey,* Nov. 23, 1973, pp. 4a and 4b.

12. The previously mentioned SPE-paper #4132, written in February/March 1972, concluded at that point that as a result, "the cheap foreign oil may not be as cheap as is being claimed, nor will its price remain at current levels, as is being claimed." Surely, no one will take exception with this statement today.

13. The cartel might even be willing to incur a temporary loss by going below its cost of production, only to reap the subsequent monopolistic profits that it will stand to gain as a result of its policy. Admittedly, this is not very likely to happen, but it is a theoretical possibility worth mentioning.

14. In fact, the cost of this research may well be borne jointly by many nations. Such a long-run policy would make a lot more sense than the hastily-proposed Organization of Petroleum Importing Countries.

15. Hartshorn, J.E.; op. cit., p. 28.

16. Adelman, M.A., "The Impact of the Tehran-Tripoli Agreements on U.S. Policy and Prices," *Journal of Petroleum Technology,* Nov. 1973, p. 1256.

17. For an assessment of the Venezuelan oil situation, see X. Boisselier and H.A.Merklein. "La Crise de l'Energie aux Etats Unis, et ses Implications sur la Politique des Compagnies Pétrolières en Amérique du Sud." A paper delivered at the Fifth Colloque Franco-Latino-Américain in Tours, France, Sept. 1973.

18. A better measure is the national income, since it incorporates the required adjustments for the replacement of capital goods and new net investment. For a detailed discussion, see H.A. Merklein, *Macroeconomics,* Wadsworth Publishing Company, 1972, Chapters 2 and 3.

19. For an excellent and fearless discussion of this problem, see W.H. Hutt, *The Strike-Threat System,* Arlington House, 1973.

20. *The Oil Daily,* Oct, 26, 1973, p. 4.

21. The elaboration that follows leans heavily on H.A. Merklein; "La Crise Energétique des Etats Unis;" *Revue d'Economie et de Gestion;* Université de Nice, France, June 1973.

22. If the price of imported oil doubled, the U.S. would have to offer (non-consume) additional items it had previously enjoyed, such as 2.3 million $4,000-cars per year. At $15 per barrel of imported oil, 8.5 million barrels per day correspond to an annual trade bill of $46.5 billion, more than 67% of the cost of all 1973 U.S. imports from all countries. If payment were to be made in $4,000-cars, the U.S. would have to export 11.6 million of them annually, more than 130% of its 1972 passenger car production.

23. For a discussion of other impending shortages, see L. Rocks and R.P. Runyon, *The Energy Crisis,* Crown Publishers, Inc., 1972, and A. Sauvy, *Croissance Zéro,* Calmann-Levy, 1973.

24. *The Wall Street Journal,* Dec. 26, 1973, p. 1.

3
Energy and Inflation

i. How Higher Oil Prices Affect U.S. Inflation

The United States is experiencing unusually high inflation rates, exceeding 12% in the Fall of 1974. Coming in the wake of the misconceived and ill-fated program of wage and price controls, two-digit inflation had cast doubt on our government's economic policies. Critics point out that the United States had been subjected to two-digit inflation only twice before in this century: during and after World War I (20% in 1917) and after World War II (14% in 1947).[1]

Since it took world wars to generate two-digit inflations in the past, it is logical to cast about for an equally massive economic disturbance that might explain today's inflation. And there just happens to be one: the 1973-1974 oil embargo and the resulting and continued high prices of imported crude oil and petroleum products.

For an apologist or an economic policy planner (these terms, alas, are synonyms these days), the increase in imported crude oil prices provides a convenient scapegoat. Because the price increase was so drastic (300 to 400%) and so rapid, and because the use of energy plays such a visible part in the production of each and every U.S. consumer good, government spokesmen have pointed to the oil embargo as a major factor contributing to the inflation problem.

But they are wrong. While recent events in the international petroleum industry have had some effect on U.S. prices, they were certainly not the major inflationary forces.

To isolate the inflationary effect of increased petroleum prices, the conventional "market basket" approach can be used. An example from a previous publication may serve to illustrate the calculation procedure used by the U.S. Department of Labor in estimating the consumer price index.[2] Basically, the consumer price index measures the cost of a given combination of consumer goods and services in one year relative to some earlier base year. This combination of goods and services is called a "market basket" and reflects U.S. consumption patterns. The price index tells the consumer how much he has to pay this year for a market basket that cost $100 last year or some other preceding base year. Table 3-1 clarifies the calculation procedure.

The market basket in column (1) of Table 3-1 contains bread, butter and milk. Column (2) shows average quantities of these goods consumed yearly by an average American family. Using the prices of the goods in the base year and in the year in question, consumption expenditures can be calculated for both years. It should be noted that quantities consumed remain constant. Furthermore, the U.S. Department of Labor goes to great trouble to make sure that the quality of the 400-odd consumer items that make up its market basket also remains the same. With quantities and qualities the same, any changes in consumption expenditures are purely a reflection of price changes.

Table 3-1
The Consumer Price Index

Market basket		Base Year		Next Year	
Good	Quantity	Price, $	Expenditures, $	Price, $	Expenditures, $
(1)	(2)	(3)	(4)	(5)	(6)
Bread	200 lbs.	.20	40.00	.25	50.00
Butter....	50 lbs.	1.00	50.00	.90	45.00
Milk	100 gals.	1.10	110.00	1.20	120.00
			Total 200.00		Total 215.00

In the table, the new market basket costs $215, while the base year's had cost $200. This can be expressed in two ways: The cost of living has risen, or the purchasing power of the dollar has been diluted. The consumer price index measures the increase in the cost of living, expressed as an index figure. In this case the price index is equal to $(215/200) \times 100 = 107.5$. Thus, the rate of inflation is said to be 7.5%.

Rather than using the consumer price index, the "GNP-deflator" has often been proposed as an inflation measure. This index is calculated exactly like the preceding consumer price index except that it includes *all* goods, not just consumer goods, since it takes into consideration the nation's entire GNP. More specifically, the GNP-deflator also evaluates the capital goods sector, the government sector, and the export-import sector in addition to the domestic consumer goods sector. In fact, inclusion of the export-import sector makes use of the GNP-deflator mandatory for a calculation of the inflationary effect of increased oil prices because the price increase originated in that sector. The "market basket" approach shown in Table 3-1 is modified for the problem at hand in Table 3-2.

As calculated in the table, the oil-related inflationary impact is equal to (1382.3/1350.0) x 100 = 102.4, or 2.4%—in other words, very little. What's more, this is a one-time impact. Once the price increase of imports and domestic crude oil has spread through the economy, leading to the above-mentioned 2.4% price-index change, there will be no further inflationary effects.

For example, another oil-induced 2.4% increase in the GNP-deflator will require that U.S. oil imports be increased again in price, say from $12.30 per barrel to something like $27 per barrel, an unlikely event. The truth is that one must look elsewhere for the dominant causes of the current price-level increase. Before this topic is taken up, however, a few annotative remarks regarding Table 3-2 are in order.

The market basket in column (1) consists of two kinds of goods: oil of various types (an unfinished good) and the remaining GNP, i.e., all of the nation's finished goods with the exception of the value added in the crude oil sector.[3] Such a breakdown is theoretically justifiable because the economic change originated in the crude oil sector. Implicit in this composition of the market basket is the assumption that, with the exception of the crude oil production phase, there have been no sudden profit increases in or outside of the oil industry. That is, profits in refining and distribution operations have not been significantly affected, nor have profits in other industries.

Price increases in crude oil are assumed to have been passed through the productive chain until they reached the final consumer in the form of increased gasoline prices, higher electricity bills, or higher prices of other consumer goods, each price-increase being proportional to (or weighted by) the crude oil value-added component of the respective good.

Another implicit assumption in these inflation calculations is that the U.S. Federal Reserve System pursues a neutral monetary policy, defined here as not changing the money supply up or down in response to the increase in crude oil prices. Ultimately, the price level always reflects the amount and velocity of money in circulation relative to the goods and services produced and sold. Thus, the Federal Reserve System could have opted for a no-inflation policy by appropriate reductions in the money supply. This would have caused non-oil-intensive goods and services to decline in price relative to oil-intensive price increases, for an overall zero inflation posture.

To the extent that price controls permitted, non-oil *energy* sectors have felt the spill-over effect of rising crude oil prices on the prices of substitute

Table 3-2
The Inflationary Impact of Crude Oil

Market Basket			Pre-Embargo Base	Post-Embargo Situation	
Good	Quantity (MMbpd)	Prices ($/bbl.)	Annual Expen. ($ billion)	Prices ($/bbl.)	Annual Expen. ($ billion)
(1)	(2)	(3)	(4)	(5)	(6)
Dom. Old Oil	6.6	3.875	9.3	5.25	12.6
Dom. New Oil	4.5	3.875	6.4	10.00	16.4
Imports	6.2	3.875	8.8	12.30	27.8
Remaining GNP ...			1325.5		1325.5
			Total 1350.0	Total 1382.3	

Notes on Table: Dom. = Domestic. Oil includes NGL production, processing gain. Old oil estimated at 60% of domestic oil production (May 1974). See *The Petroleum Situation,* June 24, 1974, p. 4, for total production and import figures, and *the Oil Daily,* June 7, 1974, p. 2 for old oil-new oil composition. See *The Oil Daily,* June 25, 1974. Supplement on Petroleum Markets and Prices, p. 1, for Pre-embargo prices. Imported oil price includes sales differential of import quota tickets. See *The Federal Reserve Bulletin,* May 1974, p. A-60, for basis of estimated GNP. Delivered cost of imported oil based on Saudi Arabian crude as listed in *Oil & Gas Journal,* May 20, 1974, p. 41. Imports also include products.

sources of energy such as coal and natural gas. However, the inflationary impact of rising prices of crude oil substitutes has been slight because:

— Price controls do exist, especially on natural gas, the second most important U.S. energy source in terms of equivalent barrels of oil.

— Oil still is the dominant U.S. energy source, representing approximately 40% of the total energy consumption.

If other non-oil price increases in the energy sector were taken into consideration, as they should be in an extended evaluation, they would be found to contribute little to the current rate of inflation.

An objection could be raised against the all-encompassing use of the item "imports," since about 45% of this was refined products at the time under investigation. Certainly, a more extensive study should subdivide imported oil not only by products but also by countries of origin. Yet, as it stands, the calculation assumes that the price of imported oil products rose in proportion to imported crude, and that is not too bad an assumption.

The quantities used in column (2) of Table 3-2 reflect May 1974 crude oil consumption data. Thus, the calculations really answer the following question: If, in May 1974, crude prices were raised from the pre-embargo level of $3.875 per barrel (June 1973), what would be the effect on the GNP-deflator? And the answer, as pointed out earlier, is that there would be a one-time 2.4% increase in that index. Yet, in the Fall of 1974, the annualized inflation rate in the United States was 12% and rising. If oil alone had been responsible, it would have been 2.4% and declining.

The 2.4% figure is certain to be doubted—even attacked—on both intuitive and empirical grounds. Intuitively, one would expect that a tripling of prices in such a large industry supplying raw materials or finished goods to practically every other industry in the United States would have ripple effects throughout the economy, severely driving up prices of all other commodities the production of which depends in one way or another on oil-related energy. That, of course, includes practically all of this nation's goods.

The answer to this line of reasoning is that prices of all other goods *are* affected, but not as much as the raw material itself, because things other than oil go into their production, and (with the exception of fuels) overwhelmingly so. In fact, to say that the inflationary impact of rising crude oil prices has been 2.4% is to say that, *on average*, all goods are affected by a 2.4% price increase.

The empirical objection, previously referred to, would be based on the observation that pre-embargo inflation rates were noticeably lower than post-embargo rates. Hence, the embargo must have been the predominant if not the sole cause of our current inflationary troubles. But empirical observations have never been very good explanatory tools in economics. In the case at hand, a number of empirical counter-observations can be cited:

1. The U.S. money supply in mid-June of 1974 was $280 billion,[4] compared to $265 billion a year earlier;[5] thus the money stock had risen by 5.7%, while the nation's real GNP had actually shown a decline.

2. Deficit spending amounted to $14 billion in fiscal 1973, with no relief in sight.[6]

3. Several rounds of labor contract negotiations had taken place and new contracts were coming up soon, with no reason to believe that unions would settle for wage increases not exceeding productivity increases.

4. Wage and price controls were lifted in April 1974, leading to a widely expected wage and price explosion from the artificially depressed earlier levels.

The United States experienced heavy inflationary pressures long before the increase in oil prices. Wage and price controls, first implemented in August 1971, preceding the oil embargo by more than two years, show that the U.S. economy was in dire straits then, as it is today. Disproportionate increases in the nation's money supply, deficit spending, and exorbitant wage settlements have been part of the U.S. economic environment in years past, they are part of it now, and will continue to be part of it for years to come, and they will perpetuate and, perhaps, accelerate inflationary pressures. So, for the dominant cause of the inflationary problem, one has to look not to OPEC but to Washington.

ii. How Higher Oil Prices Affect World Inflation

The inflation rate in the United States today is considered severe, but in other oil-importing countries—especially in Europe—it is even worse. And, in most cases, the higher crude prices instituted during the Mid-East crisis of late 1973 are cited as the major contributor to the economic problems now confronting the world. However, examination of the problem indicates that factors such as a continuing increase in the world's money supply, government deficit spending, and labor-related cost-push forces have much more influence on inflation than does the current price of oil.

The market-basket approach described in the previous section has been used as a means of calculating the inflationary impact of rising crude oil prices on various oil-importing nations. Somewhat arbitrarily, the following countries have been chosen for this study: Austria, Denmark, France, West Germany, Italy, Holland, Norway, Sweden, Switzerland, the United Kingdom, Japan and Australia. Calculation details are shown in Table 3-3.

Using the data listed or calculated in columns (1) through (9) of Table 3-3, the price-level impact due to rising prices of oil imports (both crude oil and products) was computed for each country in accordance with the following equation:

$$PI = \frac{col. (8) - col. (6)}{col. (9)} \times 100$$

In the case of Austria, for example, the calculation yielded

$$PI = \frac{0.81 - 0.20}{20.5} \times 100 = 3\%$$

This figure is shown in column (10), while column (11) shows the importance of oil imports measured in terms of each nation's GNP.

As was to be expected, the greater a nation's relative oil imports, the greater the inflationary impact. Japan's oil imports, for example, represent 1.7% of its GNP, with a corresponding inflationary impact of 5.8%. By contrast, Australia's imports are only 0.3% of its GNP, and the oil-related inflationary impact is a low 1.1%.

It is interesting to note that the oil-related inflationary impact never reaches the value of 6% in any of these countries. The average impact for all 12 countries under consideration is 4.3%.

These data are compared with the actual inflation rates of the various countries in Table 3-4 and in Figure 3-1. Column (4) in Table 3-4 shows the annualized inflation rates of the various countries based on the period December 1973 to April or May 1974, depending on availability of data. The three poorest performers in overall inflation were Japan (28.5%), the United Kingdom (23.2%) and Italy (21.3%). The United States, by the way, fared a

Table 3-3

Simulated Price Level Increase in Various Countries Due to Rising Prices of Oil Imports

	1973 Petroleum Products				Pre-Embargo Prices 1972, $/bbl.	Annual Expend. Pre-Embargo 1972, $ x 10⁹	Post-Embargo Prices 1974, $/bbl.	Annual Expend. Post-Embargo $10⁹	1972 GNP, $10⁹	Price-Level Impact (1972-Base) %	1972 Net Imports, % of GNP
	10⁶ Metric Tons per Year			10⁶ bbl. per Year							
	Cons.	Dom. Prod.	Net Imports	Net Imports							
	(1)	(2)	(3)	(4)	(5)	(6)	(7)	(8)	(9)	(10)	(11)
Austria	11.4	2.4	9.0	66	3.00	0.20	12.30	0.81	20.5	3.0	1.0
Denmark	17.8	0.2	17.6	128		0.38		1.57	21.2	5.6	1.8
France	126.0	1.3	124.7	910		2.73		11.19	195.5	4.3	1.4
West Germany	135.6	6.4	129.2	943		2.83		11.59	258.9	3.4	1.1
Italy	103.5	0.7	102.8	750		2.25		9.23	118.3	5.9	1.9
Netherlands	41.1	1.4	39.7	290		0.87		3.57	45.7	5.9	1.9
Norway	8.6	1.6	7.0	51		0.15		0.63	14.4	3.3	1.0
Sweden	28.9	...	28.9	211		0.63		2.60	41.9	4.7	1.5
Switzerland	14.9	...	14.9	100		0.33		1.34	30.8	3.3	1.1
United Kingdom	112.8	0.1	112.7	823		2.47		10.12	146.5	5.2	1.7
Japan	267.7	0.7	267.0	1949	2.60	5.07	11.60	22.61	300.1	5.8	1.7
Australia	28.4	20.7	7.7	56	2.60	0.15	11.60	0.65	45.8	1.1	0.3

Sources: Columns (1) and (2): *International Petroleum Encyclopedia*, 1974. The Petroleum Publishing Co. Columns (5) & (7): Various sources including *Platt's Oilgram*, bank releases, etc. The prices are, at best, approximate and would need finer definition in an extended study. Column (9): *International Financial Statistics*, August 1974, The International Monetary Fund.

Abbreviations: Cons. = Consumption; Dom. prod. = Domestic production; Expend. = Expenditure.

Table 3-4
Oil-Related Price Level Increases and Actual Inflation Rates

	Consumer price Index (1970 Base)			Annual Inc., Consumer Price Index, %	Oil-Related Price Level Impact, % (Table 3-3)	Inflation Due to Price Increases in Other Industries, %
	Dec. 1973	May 1974	Factor			
	(1)	(2)	(3)	(4)	(5)	(6)
Austria	124.3	129.3*	12/4	12.1	3.0	9.1
Denmark	131.2	137.4*	12/4	14.2	5.6	8.6
France	125.3	134.3	12/5	17.2	4.3	12.9
West Germany	122.9	126.7	12/5	7.4	3.4	4.0
Italy	129.4	138.6*	12/4	21.3	5.9	15.4
Netherlands	129.4	136.0	12/5	12.2	5.9	7.3
Norway	126.2	131.7	12/5	10.5	3.3	7.2
Sweden	126.0	130.0	12/5	7.6	4.7	2.9
Switzerland	131.6	133.8	12/5	4.0	3.3	0.7
United Kingdom ..	134.2	147.2	12/5	23.2	5.2	18.0
Japan	134.0	149.9	12/5	28.5	5.8	22.7
Australia..........	125.0ᴬ	132.6ᴮ	2.0	12.2	1.1	11.1

*April 1974.
ᴬ Third quarter, 1973.
ᴮ First quarter, 1974.

Sources: Columns (1) and (2): *International Financial Statistics,* August 1974, International Monetary Fund. Column (5): Table 3-3.

Abbreviations: Inc. = Increase.

little better than average in this test; its annualized inflation rate was 12.3%, compared to an overall average of 12.7% for all countries considered.

By deducting the oil-related inflationary impact (column 5) from the overall inflation rate, we find a remainder (column 6) that cannot be explained away by the oil embargo. On average, this corresponds to about two-thirds of the various overall inflation rates.

Because meaningful international data on inflationary forces are reported with substantial lags, it has not been possible to rigorously determine these forces occurring in early 1974. Yet, some insight may be gained by looking at the earlier performance of the various governments in regard to economic policies. A notoriously poor performance in earlier years may be suggestive of troubles in the most recent past. Such a comparison of economic policies is given in Table 3-5.

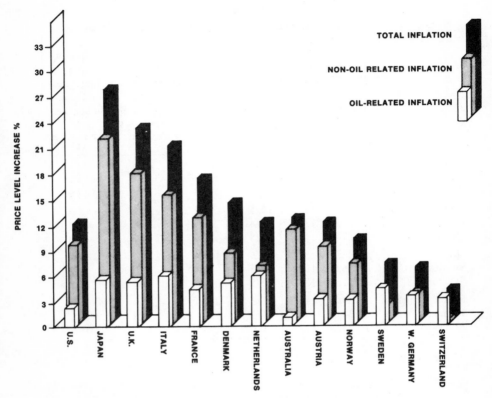

Figure 3-1. The impact of crude oil prices on inflation.

The various data in Table 3-5 are best discussed in terms of a given country; say, Italy. In the period 1970-1972, that country's real gross domestic production went up by 4.9%, while its money supply was increased by 47.5%. Thus, in accordance with column (3), Italy's money supply increased at a rate 9.7 times faster than its domestic output. Certainly, this is a poor record of monetary policies.

Columns (4) to (6) are meant to give some indication of the existence of labor-related cost-push forces in the various nations. Again, Italy fared poorly in that category: manufacturing wages increased by 55.7% at a time when prices rose by 22.4%. Thus, wage increases outran price increases by a factor of 2.5, suggesting labor problems and retarded growth rates in total production.

Deficit spending, column (7), totaled 18.5 trillion lira in 1971-73, for an annual average of 6.2 trillion lira, column (8). That was almost 9% of Italy's GNP of 1972. By comparison, the U.S. rate of deficit spending over the same period amounted to 1.75% of its GNP, column (9).

Table 3-5

Past Economic Problems and Policies

	Mid 1970-'72 Increase Real Gross Domestic Product, %	End 1970-'72 Increased Money Supply, %	Relative Money Growth, Col. 2 ÷ Col. 1	1970-'73 Inc. in Mfg. Wages, %	1970-'73 Inc. in Consumer Prices, %	Ratio Col. 4 ÷ Col. 5	Deficit Spending 1971, '72 and '73 Local Currency x 10⁹	Average Deficit Spending 1/3 of Col. 7	Deficit Spending as % of GNP
	(1)	(2)	(3)	(4)	(5)	(6)	(7)	(8)	(9)
Austria	13.9	40.5	2.9	38.2	19.7	1.9	22.2*	7.4	1.55
Denmark	8.9	33.5	3.8	49.5	23.0	2.2	(4.6)†	(1.5)†	(1.03)†
France	10.9	27.5	2.5	39.0	19.9	2.0	(6.55)†	(2.2)†	(.22)†
West Germany	5.7	28.4	5.0	36.2	18.8	1.9	6.40	2.1	0.25
Italy	4.9	47.5	9.7	55.7	22.4	2.5	18519.	6173.	8.95
Netherlands	8.5	35.3	4.2	42.0	25.2	1.7	0.8	0.3	0.20
Norway	9.9	17.2	1.7	35.0	22.0	1.6	7.2	2.4	2.50
Sweden	2.6	17.5	7.4	33.0	16.2	2.0	(3.2)†	(1.1)†	(0.58)†
Switzerland	....	24.1	....	30.0	23.6	1.3	2.3	0.8	0.69
United Kingdom ...	4.6	31.3	6.8	42.2	28.0	1.5	4.0	1.3	2.08
Japan	16.3	61.7	3.8	62.8	24.5	2.6	3463.	1154.	1.27
Australia	7.9	26.7	3.4	36.9	22.9	1.6	1.036	0.3	0.84
United States	9.2	17.1	1.9	21.0	14.4	1.5	60.6	20.2	1.75

*1969, '70 and '71.

†Surplus, 1970, '71 and '72.

Sources: Column (1): *Monthly Bulletin of Statistics*, July 1974, Vol. XXVIII, No. 7 United Nations, pp. 2, 12, 213. Column (2): United Nations, op. cit., pp. 222-224. Column (4): *Main Economic Indicators*, (OECD), May 1974.

In summary, then, the increase in the world money supply, governmental deficit spending and labor-related cost-push forces are now, and will continue to be, the predominant inflationary forces. The Arab oil embargo contributed no more than approximately one-third of the early 1974 price level increases. What is more, short of another substantial increase in crude oil prices in the near future, the oil-related price level increase will wear out, whereas the other forces will continue to assert themselves, probably with increasing severity. If this is true, then the current inflation problem is *not* fundamentally an energy problem.

If cheap energy were found in abundance tomorrow, the U.S. price level could be expected to decline only 2.4% from the current inflation rate. But from the new inflation level, the rate could be expected to continue its growth into the future unless changes are made in the growth of the money supply, deficit spending, and labor-related cost relative to productivity.

iii. How World Inflation Affects OPEC Nations

Just as rising crude oil prices affect the general price level in oil-importing nations, so do rising prices of export goods from industrialized nations affect the general price level in those countries that import from them. The OPEC countries, of course, are primary customers of industralized countries, which poses the question concerning the magnitude of the inflationary impact of rising import prices in OPEC countries. As in sections *i* and *ii*, the inflationary impact can be calculated in terms of the GNP-deflator.

Table 3-6 lists 1974 inflation rates of some selected OPEC countries. These average out at 18%, with Indonesia experiencing the highest rate at 42.9%. How much of these rates is attributable to rising world prices? Unfortunately, recent data in most OPEC countries are not easily accessible. For this reason, actual calculations in Table 3-7 were limited to three OPEC members: In-

Table 3-6
Inflation Rates in Selected OPEC Nations

Country	Consumer Price Index (1970 Base)		Annual Inflation Rate %
	Dec. 1973	June 1974	
Indonesia	168.0	204.0	42.9
Iran	128.2	141.4	20.6
Iraq	117.3	120.9	6.1
Kuwait	113.5	123.1	16.9
Venezuela	114.4	116.4	3.5
Average			18.0

donesia, Iran and Venezuela, using different world inflation rates as parameters. For example, Iran's 1972 imports amounted to 16.0% of its GNP, and a 20% world inflation rate will induce a 3.2% domestic inflation rate in Iran. If the world inflation rate rose to 30%, it would push the domestic rate from 3.2% to 4.8% over and above that attributable to purely Iranian causes.

Because Indonesia, Iran and Venezuela exhibit similar relative import volumes, a greater data spread is provided in Table 3-8, where imports were allowed to range from 10% of GNP (Country A) to 50% (Country D). As this theoretical table shows, low-import Country A is less vulnerable to world inflation than is high-import Country D. For example, at a 30% world inflation rate, A's induced inflation is 3% above its home-grown rate, while that of D equals 15%. The results of Table 3-8 are illustrated in Figure 3-2.

In reference to Tables 3-7 and 3-8, it should be noted that the GNPs as used there need to be adjusted for surplus currencies resulting from international trade. These currencies are generally held abroad and do not con-

Table 3-7
The Repercussion of World Inflation in Selected OPEC Nations

Country	1972 GNP $ x 10⁹	Imports—Base Year $ x 10⁹	Imports—Base Year % of GNP	Domestic Inflation Rates, %, at World Inflation Rates of: 10%	20%	30%	50%
	(1)	(2)	(3)	(4)	(5)	(6)	(7)
Indonesia ...	10.6	1.5	14.1	1.4	2.8	4.2	7.1
Iran	16.9	2.7	16.0	1.6	3.2	4.8	8.0
Venezuela ...	13.1	2.2	16.8	1.7	3.4	5.0	8.4

Sources: Column (1): *International Financial Statistics,* October 1974; IMF.
Column (2): *Direction of Trade, Annual 1969—73,* IMF.

Table 3-8
The Domestic Repercussions of World Inflation

Country	Imports, % of GNP	Domestic Inflation Rates, %, at World Inflation Rates of: 10%	20%	30%	50%
A	10	1	2	3	5
B	20	2	4	6	10
C	30	3	6	9	15
D	50	5	10	15	25

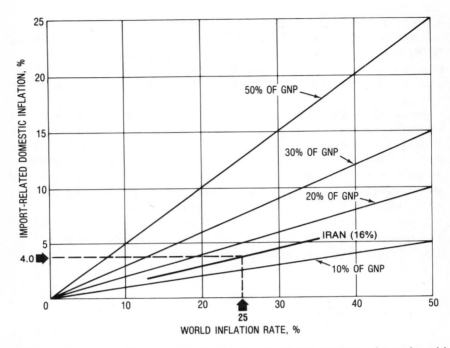

Figure 3-2. *Imported and world inflation. The domestic repercussions of world inflation can be determined for a given country from this chart. For example, suppose the world inflation rate is 25%. To determine the import-related inflation in Iran, whose imports are about 16% of its GNP, enter the abscissa at 25% and determine the intersection with Iran's 16% import line. An import-related inflation rate of 4.0% is then read from the ordinate.*

tribute to domestic inflation. The oil production that generated these surplus funds is, strictly speaking, part of the relevant OPEC country's GNP, but it has slipped out of the country without generating a reverse stream of import goods and, for that reason, it is not properly part of the inflatable OPEC GNP, nor does it contribute to inflation.

What, then, has been the trade-induced inflation rate in OPEC countries? At an assumed world inflation rate of 20%, the import-induced inflation rate in Indonesia has been about 2.8%, while the respective figures for Iran and Venezuela are 3.2% and 3.4%. Overall, the import-induced inflation rate in OPEC countries runs from 3% to 4%.

The following table shows how OPEC inflation relates to embargo-induced inflation in industrialized nations:

	Inflationary Repercussions	
	from OPEC to other countries	from other countries to OPEC
Rate	3-4%	3-4%
Type	One-shot	Continuous

To be noted is the fact that the inflationary repercussion from OPEC to other countries was brought about by a 320% increase in oil prices, while the reverse effect, essentially the same in magnitude, was the result of a 20% world inflation rate. Thus, it can be seen that OPEC countries are much more vulnerable to world inflation than the oil-importing countries are to rising oil prices. In fact, the "reverse inflation multiplier" is on the order of 10 to 15.

This vulnerability of OPEC countries to imported inflation and the concurrent financial losses incurred by OPEC are the prime reasons for their desire to adopt an indexing mechanism for crude oil prices. Without indexing, the real price of crude oil will quickly decline to the pre-embargo level of near $3 per barrel, as can be seen in Table 3-9.

Table 3-9
Effective Crude Oil Prices for Fixed Nominal Prices, $/bbl.

Year	World Inflation Rate	
	15%	20%
1	$9.35	$8.80
2	7.95	7.04
3	6.76	5.63
4	5.75	4.50
5	4.89	3.60
6	4.16	2.88
7	3.53	—
8	3.00	—

This table shows that a 15% world inflation rate will reduce the nominal crude price of $11.00 per barrel to a real price of $3 per barrel in eight years, if prices remain unindexed. At a 20% world inflation rate, pre-embargo prices will be re-established in as little as six years.

More will be said on the topic of crude oil price indexation in Chapter 6. Suffice it here to note that the failure to index the price of crude oil will give industrialized nations an incentive to inflate since this will cancel the gain that has been achieved by OPEC countries.

iv. What Subsequent Price Hikes Meant to World Inflation

There is hardly an economic issue today that is more thoroughly misunderstood than the impact of rising crude oil prices. For example, at a hearing held September 8-9, 1975, by the House Interstate and Foreign Commerce Energy-Power Subcommittee, some of the witnesses had incredible things to say about the impact of rising prices of domestic oil, i.e., of oil price decontrol.

For example, according to Ralph Nader, the Ford-Exxon oil pricing position would allow American oil producers to charge American consumers the *inflationary* OPEC monopoly price for American oil.[7] The fact of the matter is that the embargo-induced tripling of crude oil prices had at best a moderate one-time price-level impact in the United States. This has been discussed in section *i* of this chapter, where the calculations are shown and all sources referenced. Moreover, on the date of the testimony, nearly two years after the embargo, the inflationary impact of OPEC prices had long worn itself out and simply no longer existed—a dead issue if ever there was one. Nor is decontrol of domestic crude oil particularly inflationary. At the time of the testimony, prior to the 10% increase in OPEC prices, sudden decontrol and the concurrent increase in the price of old oil to the world price would have induced a one-time inflationary ripple of 1.1%.

What about unemployment? "There will be 600,000 additional people unemployed" as the result of decontrol, according to Congressman Adams (D-Wash.), citing a Data Resource Institute analysis. "It will cost about 1.4 million jobs," said Lee White, citing a study by the Consumer Federation of America. There will be "slight economic effects," according to FEA's Frank Zarb and Allan Greenspan, chairman of the Council of Economic Advisors to the President.

Obviously, not every witness can be right in the face of such conflicting testimony. And it is equally obvious that the conflicting statements can only aggravate and perpetuate the fuzzy thinking on the subject, yet there are well-established techniques to calculate the impact of decontrol. Why, then, does not everybody come up with the same answer or with reasonably close answers?

Given the confusion surrounding the impact of an increase in domestic crude prices, it is not surprising to find equally confusing and confused thinking in regard to rising OPEC prices. As a matter of historical fact, during a sometimes dramatic session in Vienna in late September 1975, the decision was taken to raise those prices by 10%, from $10.46 to $11.51 for the benchmark Arabian light crude. A few remarks concerning the decision to raise prices are in order.

Quite apart from the question of whether the move was needed or justified, it is a matter of historical fact that the most *Arab* of the participating Arab nations, Saudi Arabia, pushed for a continued freeze on crude oil prices or, at most, a token 5% increase, while Iran, a non-Arab member of OPEC, was the leader of the faction that pushed for more substantial price increases. This was duly reported in the U.S. press; after all, Sheikh Yamani's dramatic departure for London was news—big news. But few commentators dealt with the fact that here was an Arab nation displaying genuinely pro-U.S. attitudes, and fewer have thanked Saudi Arabia for it. U.S. anti-Arab feelings are deeply ingrained and carefully nursed. The man in the street in the United States has by now heard about OPEC, but he still thinks that this is an Arab institution. He doesn't know that countries such as Indonesia or Nigeria are members of OPEC, and he would be surprised if told that Venezuela was the leading force in the movement that ultimately gave birth to OPEC.

A second point that deserves mention is the fact that the OPEC price of crude oil is an *administered* price. There is no compelling reason that it should be $11.51 per barrel or $11.52 per barrel or, for that matter, $15.52. Of course, there will be certain short-term and long-term repercussions depending upon the price that is ultimately accepted, and the price ought to be set (and probably is being set) with a view to these repercussions. Yet, this is an administered price and, for that reason, is neither justifiable nor refutable on the basis of Western economic logic using the usual competitive market model.

This, too, is an area of great confusion. The OPEC argument that, in the past, great quantities of oil have slipped out of oil-producing countries at artificially low prices is true, but it is irrelevant to current pricing strategy. The pricing decision *as of today* needs to be made in light of today's market, today's non-OPEC alternatives, and today's political environment, taking into account a probabilistic assessment of future markets and environments, but never, really, the past. It is dead for OPEC, as it is for the oil-importing nations. Both sides have yet to understand this.

There is a difference between trying to justify or refute a price increase on economic grounds and explaining it. For public relations reasons, the OPEC price increase needs to be explained, and this can be easily and effectively accomplished by pointing out that OPEC is not at all pursuing a policy of raising crude oil prices. Since prices throughout the world are rising rapidly, the price of crude oil must rise at about the same rate, or else a barrel of oil will buy less and less in world markets. In other words, OPEC should gear its public relations effort in explaining what the economist would call a policy of stable or constant *real prices* which, in a generally inflationary environment, implies rising *nominal prices* of crude oil.

A more technical argument would call for constant terms of trade for OPEC oil. That topic, and a sample calculation on terms of trade, is the substance of Chapter 6. Suffice it here to say that the annual increase in export prices from industrial countries was 26.8% during the year April 1, 1974 through March 31, 1975, while oil-exporting countries raised prices by 0.8% during the same one-year period.[8] Forgetting about the early 1974 price increase and taking April 1, 1974, as a starting point on new prices, it is easy to see just from the two cited export indices that OPEC had to do something, and do it fast, to preserve the wealth-generating power of its crude oil. And, indeed, the indices suggest that 10% was not enough to recapture OPEC's terms-of-trade losses of the intervening post-embargo period.

Having said all this, what about the economic impact of the 10% increase in imported crude oil? There are two aspects to this question, the *unemployment* issue and the *inflation* issue. These will be discussed separately even though they manifest themselves jointly in the economy.

Underlying both unemployment and inflation is the issue of wealth; this has been discussed in some detail in Chapter 2 and needs no repetition here. A country, or a group of countries, experiencing deteriorating terms of trade is subject to a creeping erosion of its relative position of wealth. What OPEC tried to do at the Vienna conference was to rebalance its terms of trade by increasing the price of its dominant export commodity, crude oil.

Considering all this, and re-emphasizing that the 10% increase in world crude oil prices is insufficient to recapture the loss in wealth experienced by OPEC countries due to world inflation, it is nevertheless true that the price increase has certain economic repercussions in oil-importing countries. In terms of unemployment, what does this mean? In the absence of market rigidities in the United States, the effect on unemployment is zero. Given the market as it is, the effect is still near zero, certainly much less devastating than union-enforced wage settlements or government-enforced increases in minimum wage rates, to name two such rigidities.

What about the inflationary impact of the 10% increase in OPEC prices? Taking into consideration both the price change of imported oil and an equivalent price increase of domestic new oil, the inflationary impact is a one-time increase in the U.S. price level on the order of 0.4%. That is so low as to be statistically insignificant.Because the 0.4% inflationary impact seems incredible, the calculations are produced in Table 3-10. (See sections *i* and *ii* of this chapter for a more detailed discussion.)

As can be seen from Table 3-10, the "sales price" of the U.S. real GNP rose from \$1,433.4 billion to \$1,439.0 billion due to the 10% increase in OPEC prices and an equivalent increase in price of domestic new oil. This corresponds to an inflationary impact of $[(1,439 - 1,433.4)/1,433.4] \times 100 = 0.39\%$, or approximately 0.4%. Roughly half of this one-time inflation is attributable to the new OPEC price itself, while the other half is due to a subsequent increase in domestic new oil prices. Of course, the domestic portion, too, is indirectly the result of the new OPEC price.

To be noted is the fact that Nigerian crude prices were used in this calculation. Had the cheaper Arabian light crude been used, the result would have been virtually the same. The critical magnitude is the \$1.50/bbl. price *increase*, not the price itself. The base on which that increase asserts itself, namely total annual expenditures on oil imports, is very small in relation to the GNP and thus of no particular significance. For example, had the calcula-

Table 3-10
Inflationary Impact on the United States of the 10%
Increase in Crude Oil Prices

Market Basket				July 1975		October 1975	
Good	Quantity (MMbpd)	Prices ($/bbl.)	Annual Expend. ($billion)		Prices Prices	Annual Expend. ($billion)	
(1)	(2)	(3)	(4)		(5)	(6)	
Dom. old oil	6.3	5.25	$ 12.1		5.25	$ 12.1	
Dom. new oil	4.2	13.00	19.9		14.50	22.2	
Imports	6.1	15.00	33.4		16.50	36.7	
Remaining GNP	—	—	1,368.0		—	1,368.0	
		Total $1,433.4			Total $1,439.0		

Notes on table: Dom. = domestic. GNP = gross national product. Expend. = expenditures. Domestic oil includes natural gas liquids and processing gain. Old oil estimated at 60% of domestic oil production in July 1975. See *The Petroleum Situation,* Aug. 29, 1975, p. 4, for total production and import figures. July price of imported oil includes freight and is based on September prices of sweet Nigerian crude, landed at Gulf Coast; October price of imported oil has been estimated by adding OPEC price increase of $1.50/bo to price of landed Nigerian crude. July price of domestic new oil actually reflects average September prices with spot oil prices on the Gulf Coast quoted as high as $14/bo. October price assumes that new oil will eventually reflect the $1.50/bo OPEC price increase, even though probably later than October. See *Federal Reserve Bulletin,* August 1975, p. A-54, for basis of estimated GNP. Imports also include products.

tion been run on the basis of imported crude oil prices in the amount of $13 before and $14.50 after the Vienna conference (landed prices) the result would have been the same (0.391% as compared to 0.398%, to be exact).

To calculate the inflationary impact of the new OPEC prices in other countries, similar calculations would have to be made for each country in question. This was done for a dozen industrialized countries on the occasion of the embargo-related price increase (section *ii*). Time and space do not permit such a series of calculations here. Still, a good estimate can be developed by a short-cut method using the calculations in section *ii* and by assuming that the new inflationary impact in countries other than the United States will be in the same proportion to the 1974-impact as it was in the United States.

The 1974 inflationary impact in the United States was 2.4%—six times greater than the 1975 impact of 0.4% or, in other words, the second round of OPEC-induced inflation will be one-sixth of what it was in 1974. Using this factor one-sixth, Table 3-11 shows how the inflationary impact of the Vienna price decision affected other industrialized countries and how this compared with the concurrent overall inflation rates. A graphic illustration of the oil-induced and overall 1975 inflation rates is given in Figure 3-3.

As can be seen in Table 3-11, the OPEC-induced inflationary impact due to the 10% increase in crude oil prices is practically nonexistent. For the 13 industrialized countries under consideration, it averages out at 0.7%, compared to an average overall inflation rate of 12.5%. Actually, the 0.7% average inflation rate is somewhat overstated, because:

1. Weighting the average on the basis of actual oil import volumes would give more importance to the U.S. figure of 0.4%, since the United States is the largest oil importer in the world.

2. None of the developing countries are included, where oil imports are considerably less in relation to their GNPs, thus making the oil-related price level impacts substantially less.

In December 1976, on the occasion of the Doha Conference, crude oil prices were raised again. Saudi Arabia once more took a moderate position in the face of heavy pressure for substantial price increases. Since the proponents for high and low prices could not reach an agreement, a two-tier price

Table 3-11
Worldwide Inflationary Impact of the 10% Increase
in OPEC Crude Oil Prices

Country	Previous Price-Level Impact (1974)	Subsequent Price-Level Impact (1975)	Current Overall Inflation (June 1975)
	(1)	(2)	(3)
United States	2.4%	0.4%	9.3%
Italy	5.9	0.98	19.0
Netherlands	5.9	0.98	10.3
Japan	5.8	0.97	13.7
Denmark	5.6	0.93	10.7
United Kingdom	5.2	0.87	26.2
Sweden	4.7	0.78	10.8*
France	4.3	0.72	11.7
West Germany	3.4	0.57	6.4
Norway	3.3	0.55	12.0
Switzerland	3.3	0.55	8.0
Austria	3.0	0.50	8.4
Australia	1.1	0.18	15.7†

*May 1975.

†Weighted Average of Australia, New Zealand, South Africa.

Sources: Column (1), *World Oil*, October 1974, Table 1, column 10, in H.A. Merklein, "How higher oil prices are affecting world inflation." Column (3), *International Financial Statistics*, September 1975, p. 34.

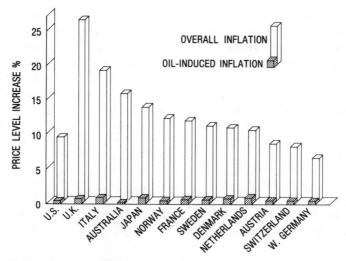

Figure 3-3. *The inflationary impact of the September 1975 10% increase in OPEC crude oil prices.*

system emerged from the Doha Conference: Saudi Arabia and the United Arab Emirates limited their price increase to 5%, while the remaining OPEC countries raised prices by 10%.

This two-tier system need not be the beginning of OPEC's break-up, as many economists have predicted. The worst that can happen, from the point of view of high-tier countries, is that Saudi Arabia cuts into their markets by increasing its production rate. This would force the other OPEC countries to accept the Saudi-UAE price. However, for reasons of resource conservation or in response to political or diplomatic pressure, Saudi Arabia may hold its production at or slightly above current levels. In refusing to relieve the world crude oil shortage, through the pursuit of a constant output policy, Saudi Arabia would create an economic environment that could support the two-tier system indefinitely.

The inflationary impact of the Doha pricing decision is roughly the same as that of September 1975: less than 1% in most oil-importing nations.

v. The Official U.S. Government Position on Oil Inflation*

One would think that the question of the inflationary impact of rising crude oil prices has now been put to rest, since most economists who have seriously

---

*This section is based on SPE Paper #6348 on "The Reliability of Econometric Models," H.A. Merklein, 7th Symposium on Petroleum Economics and Evaluation, Feb. 21-22, 1977, Dallas, Texas.

looked into the problem have come up with essentially the same answer—a rare phenomenon for this breed of people who (according to Bernard Shaw), if laid end to end, will reach no conclusion. It is generally agreed that the inflationary impact of rising crude oil prices, assuming no changes in the money supply, has been a one-time increase in the U.S. GNP deflator on the order of 2 to 3%. This has been discussed repeatedly by this author.[9, 10] Nobel Laureate Milton Friedman has been quoted as saying that "inflations are made at home by governments, and governments found it convenient to blame the oil crisis for their troubles."[11] Harvard's Professor Haberler has expressed similar thoughts,[12] as have the OECD and countless others.

The U.S. government itself has confirmed more than once that the oil-induced inflationary impact has been relatively mild. CEA's Allan Greenspan has stated that inflation is not a commodity problem, it is a monetary one.[11] In addition, the U.S. Treasury did a study on the impact of oil price increases on prices in major industrial countries.[13] On page 4 of this study, the Treasury says:

> Before the price increases of late 1973, the total U.S. crude petroleum bill amounted to roughly $22 billion, while the average of first and second quarter 1973 GNP was $1,264 billion. The total crude petroleum bill thus constituted 1.74% of GNP. After repricing at the nearly $9 per barrel average cost of crude petroleum at U.S. refinery gates, the bill is approximately $57 billion while the average of first and second quarter 1974 GNP is $1,371 billion. The crude petroleum bill now constitutes 4.15% of GNP. By this calculation, the repricing amounted to 2.41% of GNP. If the calculation is made using the $35 billion cost increase relative to the 1973 GNP figure, the percentage increase is 2.76, while if 1974 GNP is used the percentage increase is 2.55. In all of these calculations, it should be remembered that since we are dealing with current-priced GNP, some of the 1974 GNP increase is petroleum price related, thus suggesting that the upper portion of the 2.41-2.76 percentage point range is appropriate. . . . "

Apparently not content with this answer, the Treasury report then proceeds with a more elaborate input-output analysis that eventually yields answers that are more compatible with Treasury's preconceived notions, i.e., showing a higher inflationary impact.

On April 21, 1976, the U.S. Delegation to the Conference on International Economic Co-operation (CIEC), Energy Commission, submitted a report entitled: *The Impact of the 1973/1974 Oil Price Increase on the United States Economy to 1980.*[14] In that study it was claimed that, among other things, the cumulative inflationary impact through 1980 would be 9.2%, in the absence of remedial demand management measures. Coming from an agency of the U.S. government, this claim is absolutely stunning and certainly merits discussion.

The results of the Treasury study are listed in Table 3-12. Treasury's calculated inflationary impact of increased crude oil prices is plotted in Figure 3-4.

The Treasury study is weak in at least three areas:

1. It is doubtful that the inflationary impact could drag on as long as the U.S. Treasury suggests, i.e., until 1980. In fact, the only apparent reason for the inflationary impact not to go on further is that 1980 was the last year under consideration. It would be interesting to observe what the Treasury model would predict if it were run and the results listed until 1985. Moreover, the crucial years 1977-1979 are missing. Their results would certainly have helped in explaining the very odd inflation time path showing an incredible jump from 3.7% in 1976 to 9.2% in 1980, in the absence of remedial policies.

2. Given the U.S. claim that, in the absence of remedial policy measures (Table 1 of the original CIEC study), the oil-induced inflation rate will be 9.2% and unemployment 2.8%, it must surely be an acceptable proposition to all parties concerned that rising crude oil prices are a cost-push phenomenon, as perceived by oil-importing countries. Given this cost-push situation, it is clear that demand management measures, as listed in Table 2 of that study, can only do one of two things:

Table 3-12
U.S. Treasury Results Summarized
(CIEC Meeting—Paris, April 1976)

	Price of Crude Oil	GNP Deflator						
		1974	1975	1976	1977	1978	1979	1980
No Remedial Policies	Constant	114.1	122.9	130.4	M	M	M	160.0
	Up	116.4	127.5	135.2	M	M	M	174.7
	Change, %	2.0	3.7	3.7	M	M	M	9.2
With Remedial Policies	Constant	114.1	122.9	130.4	M	M	M	160.0
	Up	116.4	126.8	134.2	M	M	M	172.6
	Change, %	2.0	3.2	2.9	M	M	M	7.9
		Unemployment Rates						
No Remedial Policies	Constant	4.0	5.8	4.8	M	M	M	3.6
	Up	5.4	9.0	8.6	M	M	M	7.5
	Change, %	1.4	3.2	3.6	M	M	M	2.8
With Remedial Policies	Constant	4.0	5.8	4.8	M	M	M	3.6
	Up	5.4	8.5	7.8	M	M	M	5.1
	Change, %	1.4	2.7	3.0	M	M	M	1.5

M—Missing Data

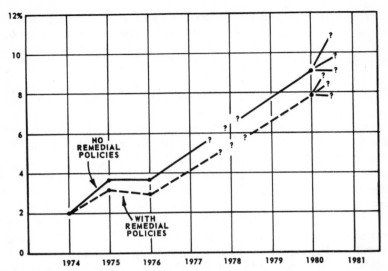

Figure 3-4. *U.S. Treasury results—CIEC meeting: The cumulative inflationary impact of rising crude oil prices (%).*

    (a) Reduce the inflation rate, but at the cost of higher unemployment.

    (b) Reduce the unemployment rate, but at the cost of higher inflation.

Yet, the model shows that improvement can be had on both fronts, a physical impossibility. Indeed, according to the U.S. position paper, management demand policies succeed in reducing the inflation differential from 9.2% to 7.9%, while simultaneously reducing the unemployment differential from 2.8% to 1.5%. This contradiction in results suggests that the Treasury model is inaccurate, its mathematical complexity notwithstanding.

    3. Finally, of course, the inflationary impact itself is much too large. Most of the difference between the generally accepted inflation figures of 2 to 3% and those of the U.S. delegation arise from imaginary "indirect" effects.

The preceding objections to the Treasury's claim regarding the impact of rising crude-oil prices are based on purely theoretical considerations. Because the Treasury based its claims on a very advanced econometric model (the Wharton model, to be exact), the question arises concerning the empirical reliability of that and of other macro-econometric models.

S.K. McKnees, in a recent article, has looked into the economic forecasting performance of five of the most widely known and influential macro-econometric models, including the Wharton model.[15] The other four models are: the U.S. Department of Commerce model, the Chase Econometrics model, the Data Resources model, and the National Bureau of Economic Research model. Examples of econometric forecasting errors from McKnees' paper are presented in Figure 3-5.

As that figure shows, the five forecasting models agreed in the first quarter of 1970 that the inflation rate a year later would fall within the narrow range of 3.0 to 4.0%. As it turned out, the inflation rate fell outside the range of high and low estimates, at 5.0%.

While the five forecasts shown in Figure 3-5 are admittedly the worst ones, the other five forecasts that are discussed in McKnees' paper fail to redeem the econometric models. For example, of the total of ten inflation forecasts discussed in McKnees' paper, there are only two instances where the observed inflation rates one year later fall within the high-low forecast brackets. In the remaining eight cases, the observed inflation rates are higher than the predicted rates, and often by a significant amount. For example, the second inflation forecast listed in Figure 3-5 ranges from a low of 3.5% to a high of 4.5%, compared to an actual rate a year later of 7.4%.

Applied to macroeconomics, econometric forecasting is not good theory and it is not good practice. In the narrow area of calculating the economic impact of rising crude oil prices, one would think that the U.S. Treasury would at least re-examine its calculations in the face of overwhelming evidence that its published data are off the mark. But the Treasury will not consider the possibility of its being in error. Given the international forum at which the referenced study was submitted, it is clear that it was in the Treasury's interest to overstate the damage that had allegedly been wrought by OPEC on the U.S. Economy and, by implication, on other oil-importing

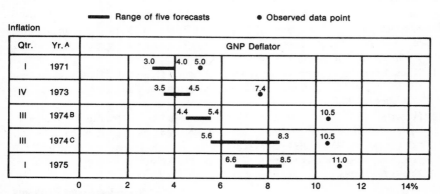

Figure 3-5. Selected econometric forecasting errors. (Source: McKnees.[15])

nations. Both from the content of the study and from treasury's reaction to reasoned objections from outside, the nagging suspicion remains that the study is more propagandistic than scientific, and deliberately so.

Be this as it may, there is a far more frightening aspect to the problem of econometric forecasting: it is the concept of econometric control.

Clearly, a policy maker who believes in the forecasting ability of econometric models has not far to go to extend this belief to the use of econometric models as a means of setting economic policy and of moving towards an econometrically controlled (that is, centrally controlled) economy. This is no longer a case of petty squabbling over the economic impact of a change in this or that internal or external variable. This involves direct governmental interference with every aspect of a nation's productive activity and, ultimately, with freedom itself.

References

1. Board of Governors of the Federal Reserve System, *Historical Chart Book, 1969,* The Federal Reserve System, Washington, D.C., pp. 96 and 97.
2. Merklein, H.A., *Macroeconomics,* 1972, Wadsworth Publishing Co., pp. 9-11.
3. For a discussion of the value-added concept see Merklein, H.A., op. cit., pp. 32 and 33.
4. *The Wall Street Journal,* July 19, 1974, p. 16, column 3.
5. Federal Reserve System, *The Federal Reserve Bulletin,* May 1974, p. A-14.
6. Federal Reserve System, op. cit., p. A-36.
7. This and the following statements were reported in *The Capital Energy Letter,* CEL 88, Sept. 15, 1975.
8. *International Financial Statistics,* International Monetary Fund, Sept. 1975, p. 32.
9. Merklein, H.A., "How Higher Oil Prices are Affecting World Inflation," *World Oil,* Oct. 1974, pp. 125-127.
10. Merklein, H.A., "The Effect of Higher Oil Prices on World Inflation," Paper No. 116 (A-1), Proceedings of the Ninth Arab Petroleum Congress, March 10-16, 1975, Dubai, United Arab Emirates.
11. Quoted from an address given by Dr. F.P. Rossi-Guerrero, Minister Counselor (Petroleum Affairs), Embassy of Venezuela, before the Fletcher Alumni Association; Dec. 9, 1976.
12. Haberler, G., "Oil, Inflation, Recession and the International Monetary System," *The Journal of Energy and Development,* Spring 1976, pp. 177-190.
13. "The Impact of Oil Price Increases on Prices in Major Industrial Countries," Office of the Assistant Secretary for International Affairs, U.S. Department of the Treasury, Research Office, Oct. 24, 1974.
14. "The Impact of the 1973/1974 Oil Price Increase on the United States Economy to 1980," U.S. Delegation to the Conference on International Economic Cooperation, Energy Commission, Paris, April 12, 1976.
15. McKnees, S.K., "The Forecasting Performance in the Early 1970's," *New England Economic Review,* July/August 1976, pp. 29-40.

4
Energy and International Money

i. The International Monetary System

In May 1971 the U.S. dollar, and with it the international monetary system, was under attack.[1] In a three-day period, holders of U.S. dollars dumped what they considered an asset of questionable value at the rate of more than $1 billion a day. U.S. oil imports, at that time, were 3.7 million bpd.[2] At a posted price of about $2.50 per barrel, this meant an annual gross outflow of $2.9 billion.

On August 15, 1971, the U.S. government abandoned its policy of keeping the U.S. dollar convertible into gold.[3] The practice had been limited in its application to official foreign holders of U.S. dollars, mostly foreign central banks. Its suspension officially confirmed what leading indicators in the international monetary markets had already pointed to: The U.S. dollar was overvalued.

Two devaluations later (in December 1971 and in February 1973), things began to look up for the U.S. dollar. It is true that the international monetary system died after the second dollar devaluation, at least insofar as its original conception was concerned, as set up in 1944 in the so-called Bretton Woods Agreement. What was originally designed to be essentially a fixed exchange rate system turned into a worldwide floating exchange rate system. Still, the

U.S. trade deficit was turned around and long-term prospects in mid-1973 were brighter than they had been in a long time, even though oil imports had risen to 5.8 million bpd.[4]

The Arab oil embargo changed all that. The immediate impact of the embargo was to strengthen the U.S. dollar. With U.S. oil imports running at about 35%, compared to an almost complete reliance on oil imports in Japan and, to a lesser degree, in most Western European countries, the U.S. economy got hurt less. In fact, the U.S. dollar rose by 13% in relation to the Japanese yen in the period November 1, 1973 to February 1, 1974.[5]

The equivalent figures for other selected hard-currency countries are listed below:

Decline in spot exchange rates against the U.S. dollar
(Nov. 1, 1973-Feb. 1, 1974)

Country	Currency	% Decline
Germany	mark	20
Switzerland	franc	10
Belgium	franc	18
Canada	dollar	0
France	franc	21
Holland	guilder	20
Italy	lira	15
England	pound	8

But the long-term impact of the embargo is different. While it appeared at first that the U.S. dollar had benefited from the oil embargo, the truth is that it got relatively less sick than the other currencies. The difference is subtle, but important. The future of both oil prices and import volumes is clouded, but one thing is certain: the post-embargo era implies U.S. and worldwide oil import bills of unprecedented magnitudes. For example, total U.S. oil imports in June 1974 amounted to 6.8 million bpd.[6] Using a conservative price of $10 per barrel for OPEC oil, this corresponds to an outflow of $68 million a day or $25 billion annually, more than four times the cost a year earlier.

Whether the current international monetary system can hold up under the changing circumstances depends largely on OPEC policy in international monetary matters. That policy, in turn, will be affected heavily by the revisions of the Bretton Woods Agreement that are being debated by the Committee of Twenty of the International Monetary Fund. The committee submitted a reform proposal in mid-June of 1974.[7] The essential innovation of the proposed international monetary system was the replacement of the dead fixed exchange rate system with what has been called a system of "managed floats." Before that system can be discussed meaningfully, it is essential to briefly outline the old system and to specifically point out where it went wrong.

The Bretton Woods System

One can objectively describe the international monetary system as it evolved from the original Bretton Woods Agreement, but one cannot objectively appraise it. A brief description will set the stage for the critique which follows in section *ii*.

The one prevailing characteristic of the international monetary system, as it operated until February 1973, was its adherence to fixed exchange rates.[8] In setting up the International Monetary Fund (IMF), the Bretton Woods Agreement did foresee the need for occasional adjustments in exchange rates, but this was thought to be a seldom used and highly exceptional emergency operation that most nations, and certainly the industrialized nations, would not have to resort to. Why fixed exchange rates? Because, so went the thinking, fixed rates facilitate international payments by eliminating re- or devaluation risks, thereby promoting international trade and investment.

International trade allows each country to specialize in the production of goods that it can produce most efficiently, due to its climate, resource base, labor force, etc. With each country producing in accordance with what has been called its comparative advantage, each country, and therefore the world as a whole, will produce more, and hence get to consume more. Therein lies the advantage of international trade.

International capital flows facilitate development of productive capacities in different parts of the world, also normally in accordance with the comparative advantage. No foreign capitalist would invest money in Saudi Arabia for the purpose of growing wheat, but billions of dollars have been invested there in the search for and development of oil production.

Fixed exchange rates, however, deprive the world monetary system of one important stabilizer: they eliminate the force that tends to keep long-term balances of payments in equilibrium. This is easily understood if exchange rates are discussed in terms of what they really are: prices of other currencies expressed in terms of a given currency.

Unbalanced money flows due to both international trade and investment are reversed by the equilibrating force of variable exchange rates. It is a well-known fact that a deficit country will be subject to downward pressures on its currency. If that pressure is allowed to assert itself, the currency of the deficit country will decline in value or float down.

In international trade the implications are clear: to the citizens of the deficit country, foreign goods become more expensive, since more of their "devalued" currency is now required to purchase the foreign currency that is needed to buy foreign-made goods. Thus, imports of the deficit country decline. By the same token, exports from the deficit country are stimulated, since the foreign currencies now buy more of the "devalued" currency. Both the reduction of imports and the stimulation of exports tend to redress the balance-of-payments deficit.

Capital flows, although in actual magnitudes considerably less significant than trade, respond in much the same way. The reduced value of a deficit currency makes investments in that country more attractive in terms of foreign currencies. Alternatively, capital flows out of the deficit country are discouraged, since investments abroad suddenly become more expensive.

This, in any event, would be the effect on capital flows, if other factors did not interfere, notably the existence of an inadequate international monetary system (the present topic) and governmental attitudes that are hostile to foreign and/or domestic capital. Two countries, such as France and Germany, that largely share the same views on capital would experience the kind of capital flow equilibration discussed above if their exchange rates were flexible.

Be this as it may, the Bretton Woods Agreement aimed at essentially fixed exchange rates. The implicit elimination of the balance-of-payments equilibrium mechanism facilitated the emergence of long-run surpluses and deficits.[9] Of course, this was recognized at the time, and much of the institutional set-up of the International Monetary Fund dealt specifically with the problem of how to overcome the anti-equilibration bias inherent in fixed exchange rates. Put differently, the post World War II international monetary system was designed in such a way as to prevent inevitable balance-of-payments deficits and surpluses from breaking up the system. This was to be accomplished by making surpluses and deficits short-lived, especially deficits. The strategy was institutionalized through an international monetary pool from which member countries could borrow funds to alleviate temporary balance deficits.[10]

Technically, the deficit country is said to "draw" on the IMF accounts. However, since payment is made with the deficit country's own currency (the country's IOU), this is strictly a loan procedure. The IMF is said to provide the liquidity required to tide the deficit country over its difficulties, thereby assuring a smooth performance of the fixed-exchange rate system. Indeed, "liquidity" became the rallying cry of international monetarists in the '50s and '60s. Much was written on the question of an optimum world-wide liquidity, since either too much or too little was deemed undesirable.

The IMF international monetary pool or general account was established in much the same way as the U.S. Federal Reserve System (Fed) was initially financed. The Fed had each participating bank make a contribution in the amount of 3% of its equity capital.[11] Similarly, the IMF had each participating country make a contribution to its general account. The amount of each country's contribution was calculated on the basis of several variables, including its international trade volume, its national income, etc. Thus, larger and more active countries contributed more than smaller countries, as large banks had contributed more to the Fed than small banks. But the comparison stops here, since:

— The Fed obtained legal jurisdiction over its founding and all subsequent member banks; it could impose binding regulations. The IMF could only consult or be consulted. Its rules could be, and frequently have been, flouted.

— Shares held by Fed member banks were non-voting, and very specific rules attempted to make sure that no bank would dominate by virtue of its size. The IMF voting rights were set in accordance with each member country's quota, thus giving larger countries (and especially the United States) a dominant role. For example, the U.S. quota was nearly 23% in June 1974, at a time when the IMF counted 126 members.[12]

The IMF general account, then, was initially established by member country quotas. In addition to the voting power, discussed above, these quotas also limited a member's use of the resources.

Member countries were originally required, at least to the extent possible, to furnish 25% of their quotas in gold and the rest in their own currencies. This was done by all developed countries and by the majority of the less developed countries. Borrowings covered by a member country's gold holdings in the Fund were referred to as *gold tranche* purchases. *Credit tranche* purchases, on the other hand, refer to borrowings exceeding a member country's gold holdings. Since these as well as gold tranche purchases are paid for (or secured by) the borrowing country's own currency, the act of borrowing will drain the Fund of borrowed currencies, usually hard currencies, and it will load it with borrowing or weak currencies. For example, on April 30, 1974, deficit countries had borrowed from the German account and withdrawn from the Fund close to 800 million SDRs worth of Deutsche marks (the SDR is a unit of account that will be defined later), or the equivalent of nearly one billion U.S. dollars. By contrast, the less developed countries taken as a group had resorted to direct net borrowings to the tune of nearly 1.6 billion SDRs, increasing the Fund's holdings of weak currencies by that amount.[13]

To prevent any one country from taking undue advantage of the IMF's borrowing facilities and thereby enabling it to perpetuate its balance-of-payments deficit, the accumulation in the Fund's General Account of a member country's currency is limited to 200% of it overall quota. However, this rule has been violated on occasion.

In addition to the relief provided deficit countries through gold tranche and credit tranche borrowings, the IMF had three other minor loan mechanisms, plus a major one, as of mid-1974.[14] Minor features were:

— *Compensatory borrowing.* A member country experiencing a drastic decline in exports could obtain hard currencies up to 50% of its quota under this provision. However, it had to accept to implement remedial domestic policies deemed advisable by the Fund. Chile, Bangladesh, the

Khmer Republic and Zambia were fully loaned out under this provision on April 30, 1974.[15]

— *Buffer stock borrowing.* A member country could use funds so borrowed to finance establishment of international buffer stocks of primary products. Only Bolivia has taken advantage of this provision, to the tune of 5.2 million SDRs, or 28% of its buffer stock ceiling.[16]

— *Standby arrangements.* These allowed a member country to implement a preconceived deficit elimination program. In return for a commitment to specific domestic remedial policies (monetary, fiscal, or others), the IMF assured the deficit country of future borrowings up to specified limits and for clearly defined periods of time.

The one major innovation in the IMF's modus operandi was, of course, the introduction of *Special Drawing Rights* or SDRs. These became legal on July 28, 1969, through appropriate amendments in the IMF Articles of Agreements, and they became operational on January 1, 1970, when participating member countries received their first allocation in the amount of 3.4 billion. Subsequent SDR allocations of 3.0 billion each took place on January 1, 1971, and on January 1, 1972.

SDRs are unconditional reserve assets that can be used by member countries to finance Fund borrowings. Since they are mere entries on the IMF's books, they have been referred to occasionally as paper gold, gold being the other unconditional, if somewhat more weighty, reserve asset.

Not all IMF member countries opted to participate in the Special Drawing Account. Interestingly enough, the Middle East oil-exporting countries (with the exception of Iran and Iraq) preferred not to join.[17]

Here is what the IMF has to say on the use of SDRs:

> SDRs may be used when a participant has a need to use reserves. When a participant wishes to use SDRs to acquire currency, the Fund may designate the participant that has to receive them and to provide the user with a currency which must be a "currency convertible in fact." Participants that are designated are those that have strong balance-of-payments and reserve positions. . . . A participant is obligated to accept SDRs on Fund designation up to the point at which its holdings are three times its (SDR) allocation.[18]

As is to be expected, hard currency countries have accumulated SDRs under this provision at the expense of less developed countries that have used them to acquire hard currencies. This is why SDRs are said to have increased world liquidity.

For example, by April 30, 1974, industrial Europe's SDR holdings exceeded its allocation by 56%, while the less developed countries only held 67% of their combined allocation. The United States and the United Kingdom, by the way, are debtors in this system, with SDR holdings of 78% and 59% of their respective quotas at that date.[19]

When the SDRs were first introduced, they were set equal in value to the U.S dollar. Through the dollar, they were tied to gold. The two U.S. dollar devaluations of December 1971 and February 1973 changed all that. By the end of 1974, one SDR was worth approximately $1.20 (U.S.). Moreover, increasing upward pressures on the price of gold made that ultimate standard of international value less and less reliable as a measuring stick for SDRs. The reform proposal of the IMF Committee of Twenty attempted to come to grips with the SDR-valuation problem, as will be seen in the next section.

This, then is a cursory description of the IMF and its handling of the international monetary system. Needless to say, an incredible amount of talent and effort went into the conception and implementation of the system. Yet, in February 1973, the international monetary system collapsed when many countries simply allowed their currencies to float. Amazingly, international trade and investment were not seriously handicapped by this development that featured the very thing the IMF had gone to all kinds of trouble to avoid: variable exchange rates.

Why did the Bretton Woods Agreement not hold up? This question will be dealt with next.

ii. Energy and International Money

An international monetary system should be designed to enhance stable economic and financial relations between nations. The Bretton Woods Agreement failed on that score because it failed to assure balance-of-payments equilibrium between trading nations. Why? For one thing, the IMF balance-of-payments objective was fundamentally in conflict with its working concept of fixed exchange rates. One cannot have fixed exchange rates and long term balance-of-payments equilibrium. Put differently, one cannot eliminate equilibrating market forces and expect, somehow, that equilibrium will still prevail. Imposition of fixed exchange rates implies elimination of the equilibrating mechanism (cf. section i, this chapter).

Fixing international exchange rates compares with fixing domestic prices through price controls. In the price-control case, the United States undertook a noble two-year experiment that thoroughly disproved its viability. It was the grandest economic experiment ever tried. It directly involved 200 million people and their productive output, a GNP of more than a trillion dollars.

Just as any economist worth his salt had predicted, the United States was quickly engulfed in an economic chaos that produced shortages everywhere: food, paper, steel, aluminum, you name it. The American public was soon disenchanted with this particular experiment, and in 1974 the whole thing was called off.

Actually, though, price controls never were really called off: the U.S. administration simply did not have the courage or the candor to admit the mistake. Controls were quietly allowed to die through nonextension of a congressional authorization for controls.

For what it's worth, only two areas remain under control: health and oil, apparently because these two are presumed to operate under a different set of economic laws. What had been proved disastrous in all other industries is obviously held to be beneficial in health and oil. This is a highly debatable point, unsupported by any theoretical or empirical (but by plenty of emotional) evidence.

Fixed exchange rates are nothing more than price controls, except that the commodities involved are foreign currencies instead of domestic goods. That is the only difference; the comparison is valid in every other respect. Just as surely as domestic price controls bring about surpluses of some goods and shortages of others, so do foreign-currency price controls (fixed exchange rates) result in surpluses of some currencies and shortages of others, called balance-of-payments disequilibria. To override this self-defeating tendency of fixed exchange rates, and for no other reason, the whole elaborate structure of the International Monetary Fund has been created.

This has resulted in all kinds of problems and inequities, such as the development of an apparent U.S. immunity to balance-of-payments deficits. For details on this issue and on the exportation of inflation, the reader is referred to Chapter 2, section *i*.

Apart from what may be called the mechanical inequities of the IMF system, there exists a very grave moral problem: a misinterpretation on the part of the IMF as to its fundamental mission.

It has been pointed out that an international monetary system should enhance stable economic and financial relations between nations. That is *all* it should do, no more. Somehow, the IMF has lost sight of this and has assumed the additional—and for a banker, the perverse—function of redistributing wealth among nations. The IMF has become less and less of a neutral international guarantor of smooth economic relations and has gradually drifted more and more into becoming a foreign aid institute, taking from the wealthy nations and giving to the less developed nations.

The foreign aid character of IMF transactions is best brought out by a numerical example. On April 30, 1974, Chile's quota in the general fund was SDR 158 million, of which one-fourth, or SDR 39.5 million, was payable and paid in gold.[20] Here is how far this gold reserve has been stretched:

SDRs—millions

1. Direct borrowing against gold 39.5
2. Credit tranche borrowing 118.5
3. Compensatory borrowing 79.0
4. SDR position 50.8

Total borrowings   287.8

Thus, putting up gold in the amount of SDR 39.5 million provided Chile with a total of SDR 287.8 million. This corresponds to an international money multiplier of 7.3. What's more, Chile had another SDR 39.5 million to go (conditionally) on buffer stock loans, plus SDR 3.8 million (unconditionally) in its SDR account. Thus, Chile's potential borrowing capacity at that time was SDR 331.1 million, the equivalent of a money multiplier of 8.4

Of course, once that limit has been reached by several developing countries, there will be demands for an increase in international credit. The usual tack is that the world is suffering from insufficient liquidity, an acceptable term, when in fact it is suffering from poverty. If the latter term were used, it would at least be clear that this is a form of foreign aid, i.e., a deliberate transfer of wealth from have-nations to have-not nations.

There is much more to be said on this question of foreign aid. A useful exercise in arithmetic would be to calculate the increase in loan availability to all developing countries through IMF alone. In mid-1974 the total IMF loan availability ran at about SDR 10 billion, or more than three times the annual U.S. foreign aid figure. Since the United States was a 23% contributor to the IMF at that time, its commitment could conceivably run to SDR 2.3 billion, roughly the equivalent of one year's national foreign aid. However, in fairness, the United States has had its own monetary troubles. As a result, it, too, is in a debtor position in relation to IMF, mostly through an SDR 500 million drawing on the SDR accounts. Industrial Europe (excluding the United Kingdom) is the real contributor to the system, in the approximate amount of SDR 2.5 billion (April 30, 1974).

A look at some of the reform proposals for the international monetary system, as put before the IMF by the so-called Committee of Twenty (C-20), confirms the growing foreign-aid involvement of the IMF.

IMF Reform Proposals

Some proposed reforms by the Committee of Twenty, especially the so-called immediate steps, warrant discussion.[21]

Among other things, the 1974 reform proposal dealt with the establishment of a permanent representative council to supervise the management and adaptation of the monetary system. Until such a council could be established, an advisory Interim Committee would take its place.

The proposal, as submitted originally, is innocuous enough. But the draft resolution recommending establishment of the Interim Committee lists an additional function not contained in the original proposal: the transfer of real resources to developing countries.[22] Again, the foreign-aid bias is very clear.

In regard to balance-of-payments adustments, the Committee of Twenty recommends the promotion of equilibrating capital flows. The disturbing point here is that many deficit countries have developed attitudes that are positively inimical towards capital. Certainly, private capital is not about to move into Peru, to use one example. That leaves public capital, or very little.

More ominous, and probably the reason member countries could not agree, is the proposal to confiscate excess reserves of surplus countries. The mechanism envisions the use of reserve indicator points. After a country has reached a certain level of these, it "would be obliged to deposit all further accruals of reserves with the Fund and to forfeit all interest earned on the deposited reserves . . . ." There is no way that all countries could ever agree to this scheme, but it was put into the Reform Outline and therefore does reflect official thinking.

A straight foreign-aid measure was the proposed establishment of a loan facility to help finance balance-of-payments deficits of developing countries that were particularly hard hit by increased oil prices. The Oil Facility has promptly been established. In its two-year life, 17 countries lent a total of SDR 7 billion to the facility. That is SDR 18 billion short of the goal of SDR 25 billion, as originally proposed by the United States which, by the way, has contributed nothing.

Surprisingly, there has been very little real change on exchange rate policies. The Committee of Twenty recommended the continuation of fixed exchange rates; even the 2¼% margin was to be retained. As a concession to existing floats, it has been recommended that the Fund authorize floating rates "in particular situations." The lack of real innovation is covered up by a new term: *managed floats.*

In addition to promoting the principle of better management of "global liquidity" (or rather loan availability), C-20 addressed itself to the sticky problem of SDR-valuation. The committee's recommendation to base the value of the SDR on a basket of 16 strong currencies was quickly adopted by the IMF. Because the constituent currencies are themselves subject to changes in valuation, the SDR will also be "floating," unless appropriate and terribly complicated adjustments are incorporated into the system. As is typical for controlled systems, it becomes more and more apparent that the simple mechanism originally held to be appropriate becomes wholly insufficient. The natural response is more controls and greater complication, with no end in sight.

On the bright side, in every sense of the word, C-20 was not unanimous on the question of establishing a link between development assistance and SDR allocations. This has been discussed before and needs no repetition here. C-20 recommended a reconsideration of the possibility and modalities of establishing such a link.

On the not-so-bright side, in every sense of the word, C-20 recommended establishment of a joint ministerial IMF-World Bank Committee to study "the transfer of real resources to developing countries and to recommend measures." Establishment of such a committee would do nothing more than to blur the lines between the development function that was originally delegated to the World Bank and the monetary function of the IMF. Had the participants in the Bretton Woods Agreement wished to merge these two functions, they could have done so in the founding process. They kept the functions separate, however. There is little evidence to support the view that things have changed so drastically in the international economic world since 1944 that what was good then is bad now and has to be eliminated. This is not to say that things are the same as they were during the Bretton Woods Conference. The tripling of oil prices has introduced abrupt changes in international money flows. Will the international monetary system, in its old or newly proposed form, be able to survive the impact of increases in world oil prices? That is the final question to be tackled here.

Oil and International Money

Two points need to be made in this discussion. The first is that, no matter what impact oil exports will have, the current monetary situation is the result of the past system. As such, certain weaknesses of this system are still very much in existence, notably the so-called 90-billion U.S. dollar overhang. Due to its liquidity, that money makes for an explosive situation if it is allowed to slush back and forth between countries.

The second point is the fundamental change the world has experienced in the wealth transfer from oil-importing to oil-exporting nations. All current and many future international monetary problems such as balance-of-payments deficits and surpluses, currency de- or revaluations, etc., are only symptoms of this fundamental change.

In absolute terms, the big oil importers suffer most: Japan, most Western European nations, the United States. But the developing countries, India, for example, are much less capable of absorbing a relatively small reduction in wealth. A millionaire will find it regrettable to lose $100,000, but he will survive; a wage earner may be wiped out financially by a $500 medical bill.

Whether the international monetary system can survive the strain placed upon it by massive money flows towards OPEC countries depends to some extent on the IMF's willingness to let the managed float be the rule rather than the exception. In view of the Fund's past preoccupation with fixed exchange rates, such a development is doubtful.

This leaves the international monetary policies of OPEC countries as the most important unknown. One thing is sure, though: it will not be in their interest to accumulate idle funds. No matter what currency is involved, be it Deutsche marks, French francs, Japanese yen or U.S. dollars, the very act of accumulation undermines the value of that currency to the detriment of its holder. Whether the oil-importing countries like it or not, the implication is that the OPEC countries, acting in their own best interest, will be limiting their oil exports to their absorptive capacities. The politically inspired oil embargo will, as a result, be followed by a partial economic embargo, acting primarily on prices rather than on volumes.

There are two practical methods that can be used to offset OPEC oil exports. The first, capital investments in oil importing countries, may or may not be left open. It has already been pointed out that many developing countries have displayed attitudes hostile to capital. Obviously, OPEC capital would not be any safer there than any other foreign capital. Indeed, one of the most important things the OPEC countries should be doing now, and they may already be doing it, is to make a global study of countries offering safe investment opportunities.

As a general rule, the industrialized nations are not prone to nationalize foreign capital. That is why they make the bulk of their foreign investment between themselves, with relatively little private capital going to developing countries.

Sooner or later (and probably sooner), all foreign oil production operations will be nationalized. Even the major oil companies have recognized the handwriting on the wall. The question remains, however, whether the various settlements are made in such a way as to assure unimpeded capital investment in the countries whose oil companies become nationalized. This is not a matter of governmental protection but of judicial action. There is plenty of evidence to suggest that mutually acceptable settlements can be reached.

Another potential obstacle to the investment of OPEC funds in industrialized nations may be the development of an anti-foreign capital bias. This may spawn the erection of barriers specifying maximum equity holdings in corporations. In fact, most OPEC countries themselves now have legislation of this type in effect.

Offsetting capital flows, then, is one method of overcoming limited absorptive capacities of OPEC nations. To the extent that those capital transfers can be implemented, they will raise the ceiling on oil exports from OPEC nations. This ceiling on oil exports is also determined by the ability of industrialized nations to formulate a policy of commodity exports to the OPEC countries. The one commodity these countries are eagerly seeking today is technological know-how. The company, or the country, that is willing to meet that need will be ahead of its competitors.

The transfer of technological know-how necessarily involves close contacts on the official level, and more importantly, on a person-to-person level; thus there is a bright spot in the future relations between oil exporters and importers. It is true that energy will no longer be cheap; but it is equally true that the OPEC countries will perceive it to be in their long-term interest to maintain good relations with the West, for the transfer of know-how is possible only in a friendly environment.

The implantation of technical manufacturing concerns in oil-exporting countries represents perhaps the most effective means of transferring technological know-how, because it involves a training ground over the entire professional spectrum: from the manual laborer (a welder, for example) to the university graduate (the chemical engineer or the manager).

Certain U.S. firms are easier to approach on such a proposition than others. Moreover, while an outright purchase of U.S. firms involves certain risks for the foreign owner, a small equity ownership, say 10 to 20% of the voting shares, is virtually risk-free, yet in certain firms this guarantees the shareholder a voice on the board of directors. These are the firms OPEC countries should seek out, especially those that are of medium size and that have a good record of innovative technological thinking.

Once represented on the board of directors, the OPEC partners have easy access to corporate presidents and they can make mutually beneficial proposals of all kinds. One such proposal might be the building, startup, and operation of a plant for a predetermined number of years and with a fixed rate of return on the investment, with a provision for gradual phasing-out of U.S. participation, say, over a five-to-seven year interval.

iii. Energy and the Floating Exchange Rate

Most central banks and international monetary agencies have consistently and intuitively clung to the belief that fixed exchange rates are absolutely essential for a stable international monetary environment. The argument was that both capital flows and trade are enhanced if the investor or trader can know at any given point in time that his claim on foreign currencies will not change in value in terms of his own currency. For example, if XYZ corporation were asked to sign a contract today to ship electronic equipment to Germany six months from now, with payment in Deutsche marks due on delivery, it would certainly help XYZ to know that this sum will at that time represent the dollar amount necessary to cover its cost and a reasonable profit. Fixed exchange rates were presumed to provide that monetary certainty to XYZ corporation and to thousands of other exporters or importers like XYZ, as well as to investors. The argument in favor of fixed exchange rates assumes implicitly that this system eliminates the risk of losses through international currency realignments, that is, it equates fixity with stability.

The international business community has consistently supported this view. The primary reason probably has been fear of the unknown: a vision that monetary chaos would result if exchange rates were left to find their own levels. In the late '60s this fear was intensified by the fact that the world was already living under a chaotic monetary regime. Repeated attacks on the U.S. dollar were followed, more often than not, by closing exchange markets, by the erection of barriers to international money flows, or by tentative, off-

and-on, individual currency floats, all of them in contravention of the International Monetary Fund's Articles of Agreement.

Floating Exchange Rates vs. Inflation

With this kind of chaos already in existence and visibly growing under fixed exchange rates, it was only natural that the business community would be suspicious of floating rates. What the business people failed to see was the fact that the international monetary jitters of the late sixties and early seventies existed because of, not in spite of, the fixed exchange rate system. Had this cause-and-effect relationship been recognized and widely accepted among business and government leaders, the floating rate would have met with considerably less resistance, and it might have come a decade earlier.

Then, too, the business community expected to be faced with the overwhelming task of having to keep up with and anticipate changes in exchange rates of a hundred or more national currencies. As it turns out, this is indeed the case for foreign investments. The typical multinational corporation is usually large enough and sophisticated enough to take this burden in stride. In fact, since exchange rates were not all that fixed under the IMF system, the multinationals have had to do this all along.

International trade, however, presents a different picture. It would be naïve to assume that the advent of floating exchange rates would make all currencies acceptable in world markets. There are a dozen, perhaps two dozen, so-called hard currencies in which the bulk of international trade is conducted. In a way, the Peruvian sol/U.S. dollar exchange rate, or fluctuations in that rate, are irrelevant to world trade. Peruvian importers must tender hard currencies in world markets to finance their imports, and these currencies come either from exports or from bilateral or multilateral credits as they are typically granted under foreign aid or through the IMF. Because soft currencies are not part of the world trade mechanism, variations in their exchange rates have little or no impact on world trade.

A relatively new concept today asserts that one of the disturbing features of floating exchange rates is that they themselves contribute to inflation. The argument runs as follows: If a country's currency floats down, its goods become cheaper in world markets. This causes other countries to step up imports, thereby adding to the domestic aggregate demand and setting up additional demand-pull pressures on prices in the country whose currency has floated down.

The argument is fallacious. A country's currency floats down only if it is overpriced, and it is overpriced because its central bank has printed too much of it in relation to the country's real GNP. In fact, one of the virtues of floating exchange rates is that the act of eroding the domestic purchasing power of a country's currency automatically erodes its international purchasing power. Under floating exchange rates, there is no such thing as a currency exhibiting strength at home and weakness internationally, or vice versa. Mistaken monetary policies catch up with the issuing country both at home and abroad, and just about simultaneously. To see this, one would have to ask why the currency floated down to begin with. Whatever the reason, clearly, the floating down of a country's currency is a reaction to a previous disturbance, and the float-equals-inflation argument ignores this disturbance.

The fact is that trade has grown in real terms under the floating exchange rate, thereby dispelling once and for all the myth that fixed rates are essential for the conduct of international business. More importantly, the period following February 13, 1973, was marked by great monetary stability, even though individual currencies have been adjusted up or down, notably the British pound, which dropped below the magic two-dollar floor in early March 1976. Six months later it had declined to $1.63/£—an 18½% downward float. But there have not been any massive money transfers triggered by speculation over impending devaluations of this or that currency.

The United Kingdom may serve as an example in support of the argument that floating rates do not induce inflation. The dismal monetary performance of the United Kingdom has been mentioned in Chapter 3, in the discussion of factors other than crude-oil prices that contribute to inflation rates. It was mentioned at the time, (and shown in Table 3-5), that its increase in money exceeded the increase of its real gross domestic product in the period 1970-1972 by a factor of 6.8, compared to 1.9 for the United States. This performance does not rank the United Kingdom among the more conservative governments of industrialized nations, nor does its monetary performance after 1972. Comparing its quarter by quarter performance in the period 1973 through 1975, here is what emerges: the United Kingdom's real GNP remained static throughout this period at about £14 billion per quarter, based on 1970 prices, while its money stock was increased by approximately 42%, from £12.3 billion to £17.5 billion. This relentless printing of money is what caused the United Kingdom's extraordinary 1975 inflation rate of 25% and, under floating exchange rates, the continued downward float of its currency. The float is a reaction, not a contributing factor, to this inflationary process in the United Kingdom, as it is elsewhere.

How the Floating System Protects International Money Markets

One of the remarkable monetary phenomena of recent times is the fact that the world monetary system has smoothly absorbed the crude-oil price increase of 1973-1974. It has done so largely because the floating system was already in existence. Because the United States was a relatively small importer of crude oil (even though in absolute terms it was then and is today the largest oil importer in the world), the monetary market reacted by an upward float of the U.S. dollar vis-à-vis most European currencies as well as the

Japanese yen and others; and that was all there was to it. If the oil embargo and the tripling of crude oil prices had occurred under the old system of fixed exchange rates, the result would have been widespread chaos. The market would have recognized that the new oil prices would put a premium on U.S. dollars, and speculative dumping of hard currencies for the purpose of buying U.S. dollars would undoubtedly have closed down the international monetary market and might have spawned the erection of new barriers to capital flows across international borders. There would have been de- and revaluations on a hit-and-miss basis, with speculative money transfers likely to overshoot the new equilibrium, thus causing reverse speculation and more uncertainty in the markets.

The remarkable aspect of the orderly market reaction to rising crude oil prices is that no one had to worry about new parities. They emerged all by themselves, and they were automatically correct, certainly more reflective of the new economic environment than a set of government-imposed parities could have been.

If the increase in crude oil prices had taken place under fixed exchange rates, the ensuing turbulence in international monetary markets would have been blamed on the OPEC countries. Since the increase came under floating rates, there was no turbulence and there was no one to blame. Thus, the existence of floating rates turns out to have worked in OPEC's favor, for it is questionable that the OPEC countries would have recognized the structure of the monetary system as the real culprit, and if they had, no one would have been prepared to believe it.

Other shock factors besides crude oil price increases can disturb the international monetary situation: strikes, violence, political upheavals, to name a few of those that strike with lightning speed; but there are others, more gradual but in the long run equally disruptive, such as unlimited money printing or boundless deficit spending. None of these factors can be as disruptive under floating exchange rates as they are under fixed rates.

Contrary to generally accepted views, floating exchange rates isolate a country from the world. Policy mistakes, such as the United Kingdom's overliberal printing of money, come back to haunt the country of origin rather than being dispersed to other countries, as they would be under fixed exchange rates. Inflation can no longer be exported as it was by the United States in the '60s; it stays at home, causing people visibly and personally to suffer the consequences of the economic policies of their own governments. Perhaps this is one of the reasons most governments dislike floating exchange rates. In an elective political system it is likely that governments fear they will be held accountable for their policy mistakes, not so much because mistakes were made but because they were made transparent. However, such fears are unfounded, as there is little hope that the electorate will ever become sophisticated enough in its economic thinking to lay the blame for mistaken domestic policies—and especially for international repercussions of such policies—where it belongs.

The IMF as Welfare Bureau

The International Monetary Fund recognizes that the old Bretton Woods System is dead. Rather than embracing a system of floating exchange rates, the IMF takes the position that the world is currently going through a period of transition and that a new monetary system will eventually emerge that will certainly contain an element of exchange rate rigidity. Some of the IMF's reform proposals have been discussed in the previous section. Let us now discuss the newest developments.

While an international monetary system featuring floating exchange rates enhances the stability of international monetary markets and therefore fosters worldwide trade and investment, it is clear that such a system benefits the developing countries only in an indirect and somewhat diffuse way. Poverty, not monetary instability, is the number one problem of developing countries. And poverty, i.e. insufficient productive capacity, cannot be eliminated by the implementation of a more efficient or more stable international monetary system. The IMF is well aware of this, and for this reason it has assumed the role of an international redistributor of wealth whose stated policy objective includes the transfer of real resources to developing countries. Since the emerging monetary system cannot accomplish this objective by itself, the answer has been to extend the IMF's credit facilities to developing nations.

Some of these extensions have already been discussed in the preceding two sections, notably the compensatory financing facility that was established in 1963, the buffer stock facility (1969), the SDR facility (1970), and the oil facility (August 1974). Because each facility (with the exception of the short-lived oil facility) is subject to a fixed ceiling for each member country, many of the developing countries soon began to crowd against the various ceilings, and the pressure was on for additional credit facilities or for the extension of existing facilities. These were established as shown below.

1. The Extended Fund Facility, established in September 1974, accomplished two objectives: It increased the loan availability of member countries and liberalized it by providing substantially longer maturities.

2. On December 24, 1975, the compensatory financing facility was raised from 50% of each country's quota to 75%. To make this extension workable, the overall ceiling on the Fund's holdings of a member country's currency (200% of its quota, see section *i* of this chapter) was set aside.

3. To relieve the pressure of loan demands which, inter alias, was aggravated by the expected termination in early 1976 of the oil facility, and

to provide an orderly transition to a proposed increase in quotas then under debate, the IMF increased the loan availability to member countries under the Fund's normal tranche policies by 45%, i.e., from 25% to 36.25%. This policy became effective on January 20, 1976.

4. In March 1976, the Board of Governors of the IMF approved an increase in the quotas of member countries by 33%, from SDR 29.2 billion to SDR 39 billion, subject to the acceptance of an appropriate amendment of the IMF's Articles of Agreement. The effect of the quota increase is not only an across-the-board increase of credit availability in the general account. Since other credit facilities such as the compensatory and buffer stock accounts have their individual credit ceilings expressed in percentages of member countries' quotas, an increase in these quotas automatically extends the loan availability under these and other facilities. Because legislative approval of member countries is generally needed for IMF amendments, the quota increase has not taken effect by press time, January, 1977.

5. An additional and perhaps minor credit facility was created on May 6, 1976, when the IMF established its so-called Trust Fund. Receiving its funding from profits obtained by auctioning off certain gold holdings, and supplemented by voluntary contributions or loans from wealthy member countries, the Trust Fund is scheduled to provide special assistance to member countries with very low annual per capita incomes, SDR 300 or less.

Much more could be said to describe the IMF's preoccupation with wealth transfers, but the trend is clear enough: the International Monetary Fund behaves more and more like a welfare institution and, for that reason, less and less like a monetary agency. If this trend continues, there will be no end to the creation of unsecured credits, and that, in and of itself, is inflationary. If the recent acceleration of credit creation maintains its momentum, the resulting credit explosion may well lead to world-wide hyperinflation and to a subsequent collapse of all faith in any kind of paper, money or credit. That is, the IMF, which is committed to enhancing stability in world monetary markets, is now pursuing a policy that disrupts that stability.

The IMF's proposed second amendment to the Articles of Agreement, accepted by the Board of Governors in early May 1976, subject to acceptance by the members of the Fund, conveys some idea of the IMF's long-term objectives. The proposal contains the following major provisions:

1. The possible adoption of stable (pronounce fixed) but adjustable par values of member currencies. However, because some members clearly intend never again to be subjected to fixed exchange rates, participation in that system is to be a matter of choice for each member country;

2. The elimination of the gold tranche (henceforth to be called reserve tranche) and generally a de-emphasis of gold: abolition of the official gold price, the eventual sale of the Fund's gold holdings, etc;

3. Replacing gold with the Special Drawing Right as the principal reserve asset of the international monetary system and elimination of the need requirement for loans on SDR accounts;

4. Various procedural and structural simplifications.

The long-run trend, clearly, is for more and more paper assets, for an explicit preoccupation with the welfare of nations, and for an eventual merger in operations with the World Bank, created by the Bretton Woods Agreement for the express purpose of providing loans to developing countries. And it should be said in passing that the World Bank has used a more cautious and business-like approach in providing these loans than the IMF is currently taking.

A system of flexible exchange rates provides greater resilience and therefore better prospects for stability in international money markets. Being a nonintervention system, it requires practically no supervision or administration and thus comes considerably cheaper. Just as unregulated domestic prices make for undistorted markets within a given country, so do unregulated exchange rates minimize distortions in international markets.

Why the floating rate is not now universally accepted is something of a mystery. It has already proven itself by absorbing the international monetary shocks of tripled crude oil prices while at the same time returning a turbulent and totally unpredictable monetary market to sanity. In the process the floating exchange rate system has acquired empirical validity; it is no longer a theoretical curiosity. There is no reason to believe that floating exchange rates will not perform in the future as they have in the recent past, and that, certainly, is a convincing recommendation.

References

1. Federal Reserve Bank of Chicago, International Letter, Number 12, May 7, 1971.
2. Survey of Current Business, U.S. Department of Commerce, December 1971, Vol. 51, Number 12, p. S-35.
3. Nation-wide address by President Nixon, carried live on major television and radio networks on Aug. 15, 1971.
4. Federal Reserve Bank of Chicago, International Letter, Number 128, July 27, 1973.
5. Federal Reserve Bank of Chicago, International Letter, Number 157, Feb. 15, 1974.
6. "Industry Score Board," *Oil and Gas Journal*; July 8, 1974.

7. "Outline of Reform," IMF Survey, June 17, 1974, pp. 193-208.
8. United Nations Monetary and Financial Conference, Bretton Woods, N.H., July 1-22, 1944, Final Act and Related Documents, pp. 31 & 32.
9. There are a great many sources for balance-of-payments data. The most authoritative is the series entitled *International Financial Statistics,* put out monthly by the IMF in volumes exceeding 400 pages. The series will henceforth be referred to as IFS.
10. United Nations Monetary and Financial Conference, op. cit., Annex A: Articles of Agreement, pp. 28-67.
11. The Federal Reserve System, Purposes and Functions, Board of Governors of the FRS, Washington, D.C., 1963, p. 26.
12. *IFS,* June 1974.
13. *IFS,* June 1974, p. 8.
14. *IFS,* June 1974, pp. 6, 14 and 15.
15. *IFS,* June 1974, pp. 8 and 9.
16. *IFS,* June 1974, p. 8.
17. *IFS,* June 1974, p. 7.
18. *IFS,* June 1974, p. 6.
19. *IFS,* June 1974, p. 7.
20. These and all subsequent numerical data in this section are from *IFS,* op. cit., June 1974, pp. 6-16 and pp. 415-416.
21. Except where stated otherwise, the source of this section is the "Outline of Reform," IMF Survey, June 17, 1974, pp. 193-208.
22. IMF Survey, June 17, 1974, p. 179.

5
Energy and OPEC Wealth

i. Mid-East Petrodollars: Enough To Buy Corporate America?

We have seen in Chapter 3 that the inflationary impact, both in the United States and worldwide, of rising crude oil prices is considerably milder than is commonly believed. The public may be forgiven for not knowing the facts: It does not possess the technical training to check the accuracy of official statements claiming that our current inflation problem is oil-induced.

For example, in September of 1974 the U.S. Department of the Treasury underlined the predominant role oil had presumably played in pushing our inflation rate into the two-digit bracket.[1]

Slowly, very slowly, and mostly by the force of events, the notion is sinking in that the real oil problem facing the United States and other oil-importing nations is a transfer of wealth to oil-exporting nations. This was pointed out in Chapter 2, but the message bears repeating.[2]

Of course, no one will dispute the point that oil-exporting nations are wealthier now than they were in early 1973, when crude oil sold for around $3 per barrel.[3] What people in oil-importing countries find difficult to see is that their own wealth has been reduced. Yet the first statement implies the second, since the combined wealth of all nations has not been appreciably affected: the global product is still pretty much the same as it was before the embargo; only the claims on that production have shifted, in favor of oil-exporting nations.

Just how wealthy are the major oil-exporting countries? Listed in Table 5-1 and shown in Figure 5-1 are the 1974 production rates of Middle East oil-producing countries, the only countries that will be considered here. After adjustment for domestic consumption, column (4) in Table 5-1 shows annual production, by country, available for exports. Because this column will be the basis for all subsequent wealth calculations, it deserves a few qualifying remarks.

First, individual country totals may not warrant the extrapolation that will be attempted here. This may be because some of the Middle Eastern countries may opt to withhold production in an attempt to prolong the remaining life of their reserves, or it may be that some countries will be experiencing a physical decline in productive capacity. Be this as it may, the region as a whole can easily maintain an export rate of 20 million bpd (7.44 billion bbl/yr.) over the next 10 years, since Saudi Arabia and Iran, presently accounting for two-thirds of the total Middle East oil production, have a virtually unlimited capacity to step up production.

Table 5-1
Middle East Oil Production
1974

	1974 Daily Production (thousand bbl.)	Annual Production (10^9 bbl.)	Annual Consumption (10^9 bbl.)	Production Available for Exports (10^9 bbl./yr.)
	(1)	(2)	(3)	(4)
Saudi Arabia.......	8,210	2.99	0.09	2.90
Kuwait.............	2,276	.83	0.04	0.79
Iraq...............	1,850	.67	0.03	0.64
Abu Dhabi.........	1,411	.52	0.01*	0.51
Neutral Zone	541	.20	0.01	0.19
Qatar	510	.19	0.01*	0.18
Oman	291	.11	0.01*	0.10
Dubai	242	.09	0.01*	0.08
Bahrain	67	.02	0.02	—
Total Arab Middle East	15,398	5.62		5.39
Iran	6,022	2.20	0.15	2.05
Total Middle East ..	21,420	7.82		7.44

*Estimated
Sources: Column (1)—*World Oil,* August 15, 1976, p. 42.
            Column (3)—*International Petroleum Encyclopedia,*
            The Petroleum Publishing Co., 1974, p. 343.

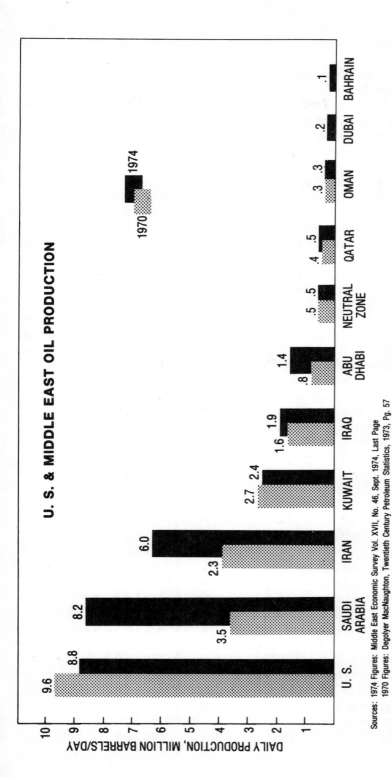

U. S. & MIDDLE EAST OIL PRODUCTION

DAILY PRODUCTION, MILLION BARRELS/DAY

	U. S.	SAUDI ARABIA	IRAN	KUWAIT	IRAQ	ABU DHABI	NEUTRAL ZONE	QATAR	OMAN	DUBAI	BAHRAIN
1974	8.8	8.2	6.0	2.4	1.9	1.4	.5	.5	.3	.2	.1
1970	9.6	3.5	2.3	2.7	1.6	.8	.5	.4	.3		

Sources: 1974 Figures: Middle East Economic Survey Vol. XVII, No. 46, Sept. 1974, Last Page
1970 Figures: Degolyer MacNaughton, Twentieth Century Petroleum Statistics, 1973, Pg. 57

Figure 5-1. How oil production is changing in the United States and the Middle East. Output boosts since 1970 in Saudi Arabia and Iran alone are almost equivalent to current U.S. production. Neither Saudi Arabia nor Kuwait are producing at capacity. (Sources: 1974 figures—World Oil, August 15, 1976, p. 42; 1970 figures—Degolyer MacNaughton, Twentieth Century Petroleum Statistics, 1973, p. 57.)

But many questions still loom on the demand side. Oil developments elsewhere (the North Sea, the Arctic Slope, U.S. offshore activities, etc.) as well as the development of alternative energy sources (coal, oil shale, nuclear power, etc.) all affect the demand for Middle Eastern oil. Moreover, the long-run consumption response to high oil prices in importing nations is an important variable, even though its impact may not be fully resolved for years. Early signs point to a slackening of import demands, judging from a 10% decline in Middle Eastern oil production in 1975.

In this section, no attempt will be made to deal with the question of demand for Middle Eastern oil; thus the wealth discussed herein is based solely on supply considerations and, therefore, is to be used with caution. It is more in the nature of potential rather than real wealth. Yet, as things are now, policy reactions by oil-importing nations on alternative sources of fuel can be counted on to be slow. For this reason, the wealth projections will be reasonably accurate for the next few years.

A second qualification involves the increasing domestic energy consumption of the oil-exporting nations as they expand their own industrial capacities and infrastructures. The projections to be used here are based on export availability. Because they hold exports constant, the calculations assume by implication that OPEC's rising domestic consumption demands will be met by stepping up production for that purpose. Again, there is no question that this can be accomplished for the region as a whole.

The third and last qualification deals with the price variable and OPEC's take, those other determinants, besides production, of the oil-exporting nations' wealth. OPEC's 1975 average take of Middle Eastern crude oil has been estimated at $10 per barrel, given the price of $10.46 that prevailed in the period January through September 1975, and $11.51 during the remainder of the year.

No one can tell what OPEC's crude oil prices will be in the future. However, the inclusion of inflation escalators in the price of OPEC's marker crude and the resultant indexation of world crude oil prices would assure that these prices will remain constant, in terms of 1975 dollars. Therefore, the following comparison with current U.S. corporate equities is reasonably valid.

The 1974 annual export rates and the assumed OPEC take of $10 per barrel of oil yield the net revenue figures shown in Table 5-2. For example, Saudi Arabia's 1974 net oil revenue was calculated to be $29.0 billion. This corresponds roughly to 54% of the net oil revenues of the Arab Middle East, or 39% of total Middle East oil revenues.

Contrasting Saudi Arabia's $29.0 billion annual revenue with its $12 billion annual development plan financed by oil revenues, it turns out that the country has 17.0 billion petrodollars in uncommitted funds.[4] These are funds available for imports of consumer goods and for investments abroad.

To arrive at comparable figures in other Middle Eastern countries, it was assumed that those countries would set aside equivalent portions of their total

Table 5-2
1974 Middle East Oil Revenues

	Production Available for Exports 10 bbl./yr.	Assumed 1975 Take $/bbl.	Net Oil Revenues			Domestic & Development Funds			Uncommitted Oil Revenue		
			Per Day $ x 10	Per Month $ x 10	Per Year $ x 10	Per Day $ x 10	Per Month $ x 10	Per Year $ x 10	Per Day $ x 10	Per Month $ x 10	Per Year $ x 10
	(1)	(2)	(3)	(4)	(5)	(6)	(7)	(8)	(9)	(10)	(11)
Saudi Arabia.........	2.90	10.00	79	2.42	29.0	32	1.00	12.0	47	1.42	17.0
Total Arab Middle East .	5.39	10.00	148	4.99	53.9	61	1.87	22.4	87	2.62	31.5
Total Middle East........	7.44	10.00	204	6.20	74.4	85	2.58	31.0	119	3.62	43.4

Source: Column (7), Saudi Arabia—Saudi Arabian five-year development plan as reported in *Middle East Economic Survey*, Vol. XVII, No. 46, Sept. 6, 1974, p. 4.

oil revenues for purposes of domestic development. This assumption yields a total of 31.5 billion petrodollars annually in uncommitted funds for the Arab Peninsula, and 43.4 billion petrodollars for the entire Middle East, here defined as Saudi Arabia, Iran, Kuwait, Iraq, Abu Dhabi, Neutral Zone, Qatar, Oman, Dubai and Bahrain.

Just how much wealth does $43.4 billion represent? If U.S. corporations could be purchased for the book value of their stockholders' equities (defined as capital stock, surplus, and retained earnings), the sum of $43.4 billion would buy the following corporations:[5]

Company	Stockholders' Equity (billions)
General Motors	$12.6
Exxon	13.7
Ford Motor Co.	6.4
Chrysler	2.7
General Electric, and	3.4
58% of Texaco	4.6
	$43.4

Going down the *Fortune* 500 List, uncommitted funds from the Middle East available in the second year would buy the stockholders' equities of the following corporations: the remainder of Texaco, Mobil Oil, IBM, IT&T, Gulf Oil, Standard Oil of California, Western Electric, U.S. Steel, Westinghouse Electric and part of Standard Oil of Indiana.

The third year's equivalence of Middle Eastern uncommitted funds would include the remainder of Standard Oil of Indiana, E.I. du Pont de Nemours, General Telephone & Electronics, Shell Oil, Goodyear Tire & Rubber, RCA, Continental Oil, International Harvester, LTV, Bethlehem Steel, Eastman Kodak, Atlantic Richfield, Esmark, Union Carbide, Tenneco, Procter & Gamble, Kraftco, Greyhound, Boeing, Caterpillar Tractor, Rockwell International, Occidental Petroleum, Firestone Tire & Rubber, Dow Chemical, McDonnell Douglas, plus one-half of Phillips Petroleum.

The fourth year would buy the next 48 largest corporations, down to number 87, American Home Products (New York). The fifth year would yield an additional 76 U.S. corporations, and the entire *Fortune* 500 corporations would change hands after 6½ years.

Of course, things are not all that bad, and for several reasons:

1. Two major classes of import goods will also be financed through these uncommitted OPEC funds: consumer goods and military hardware. Some effort should be made to arrive at a good estimate of these, but it will suffice here to say that they will be on the order of 10% each of the uncommitted funds, with some OPEC countries spending considerably more on defense goods.

2. The preceding discussion assumes that the entire Middle East will use all of its uncommitted funds to purchase corporate equities in the United States only. This, of course, is a spurious assumption. Such a course of events would completely destroy the international monetary markets, strengthening the U.S. dollar in the process and pulling the rug out from under the currencies of all other oil-importing nations.

A more realistic assumption would be to base reinvestment on each country's import bill. For example, U.S. imports of petroleum and petroleum products averaged 6.0 million bpd in both 1974 and 1975.[6] Products comprised about 30-40% of this total, but since OPEC derives its economic rent exclusively from its production operations, its crude oil take of $10 per barrel still applies. Thus, U.S. oil imports of 6.0 million bpd generate a net revenue stream to oil exporters of some $60 million a day, or $22 billion a year.

The Saudi Arabian development fund runs approximately 36% of that country's total oil revenues. Applying that factor to all U.S.-generated oil revenues leaves $14 billion of yearly uncommitted funds. Setting aside an assumed 10% each for domestic consumption and military hardware to be provided from U.S. sources, approximately $11.2 billion petrodollars remain annually for investment in the United States.

It was mentioned earlier that the combined Middle Eastern uncommitted oil revenue would buy up the stockholders' equities of the *Fortune* 500 corporations in 6½ years. The purchasing power at work in this case was $43.4 billion per year. Discarding the sinister prospect of a joint acquisition of corporate America by the combined purchasing power of all oil-exporting nations, and basing reinvestment calculations on the value of U.S. oil imports only, the resulting annual purchasing power of $11.2 billion represents 26% of the combined figure. This means that the *Fortune* 500 corporations could be bought up in about 26 years rather than in 6½, if outright ownership is the target of the oil-exporting countries. If they are content with corporate control, say through a 50% ownership of corporate equities, the corporations can be had in 13 years.

The foregoing calculations depict OPEC's ability to purchase U.S. corporations on the assumption that the stockholders' equities reflect the market values of these corporations. This is, of course, not the case. The stockholders' equities reflect corporate book values net of liabilities, rather than replacement or acquisition costs. Accordingly, the calculations understate the value of the *Fortune* 500 corporations by a considerable amount. Moreover, the unleashing of unrestrained OPEC purchasing power in U.S. equity markets would quickly drive up the market value of corporate shares to a multiple of current prices. Thus, the transfer of U.S. corporations to oil-exporting nations is not likely to happen in 26 years, and not even in 50 years. In addition, portfolio considerations by foreign governments, U.S. antitrust laws, protective legislative moves by the U.S. Congress, and other factors will be certain to prevent such a concentration of corporate equities in foreign hands, should OPEC move in that direction.

Still, the enumeration of purchasable U.S. corporations serves to illustrate the point that the wealth transfer is substantial, if looked at in absolute terms. Considered in comparison with other wealth-consuming items, it is not all that much out of line. For example, the planned U.S. defense budget for fiscal 1975 ran at $81 billion, or more than three times the concurrent outlay for oil imports. Total U.S. transfer payments amounted to $160 billion in 1975. Put differently, the net revenue accruing to exporting nations from U.S. oil imports, $22 billion per year, represented no more and no less than 1.5% of the concurrent U.S. GNP.

Early post-embargo statements by highly placed U.S. government officials were worded so as to permit the interpretation of armed outside intervention. President Ford, in an address to the International Energy Conference in Detroit, Mich., on September 23, 1974, spoke of a threatened "breakdown of world order and safety." Secretary of State Kissinger warned the U.N. General Assembly on that same day that "it is no longer possible to imagine that conflicts, weapons and recession will not spread." Stripped of all oratory, what does this mean? It means that the United States was prepared, or wished to appear to be prepared, to consider the use of force to prevent oil exporters from getting 1.5% of its GNP.

If we as a nation are as serious about the energy problem as we say we are, why don't we incur the self-sacrifice that is needed to get us past the fossil-fuel energy age? To say that there is an apple shortage is the equivalent of saying that apples are underpriced. And if the price is the result of government-imposed ceilings, then the shortage is a creation of the government. Similarly, the energy shortage means nothing more or less than that energy is underpriced, and it, too, is government-created or, at the very least, tolerated.

Let gasoline and electricity and other forms of energy double or triple in price. If that happens, we won't need legislation requiring that thermostats be set at 80°F in the summer and 60°F in the winter, or that we must form car pools. Let the price of energy quadruple, and every energy-consuming U.S. citizen will become a conservationist. What's more, the needed funds for the development of alternate fuels will be made available, through profits.

That is the kind of self-sacrifice that is needed. Are we really willing to accept it? Of course not. As always, we look for the easy way out, and that includes—or at one time in the recent past included—saber rattling by the U.S. government, that is, the American people. It is interesting to note that all the European oil importers conspicuously disassociated themselves from the United States' belligerent position, even though they were hurt considerably more by the increase in crude oil prices.

Does this mean that oil-exporting nations should be intransigent in their position? Of course not, but we must understand that we would be very likely to behave the way they do if we were in their shoes. Most of them are faced with the once-in-a-nation's-lifetime opportunity to acquire wealth over the short span of two to three decades. What they don't get now, they'll never get. That's the truth, and they know it. Yet, by a judicious selection of their investment opportunities, the OPEC nations can blunt the economic impact of their policies in oil-importing countries. This subject will be taken up next.

ii. How Middle East Producers Can Preserve Their Petro-Wealth

The potential wealth of Middle Eastern petroleum-exporting countries is enormous—indeed, nearly incomprehensible. But unless it is carefully administered, this wealth may be exhausted. If this were to happen, it would return the people in this area to relative wealth status similar to the one they experienced before the emergence of oil as a major world resource. History abounds with examples of peoples who, for one reason or another, had enjoyed a brief span of extreme wealth, only to slide back to obscurity.

Preservation of their current oil wealth is, and ought to be, the chief concern of Middle Eastern oil-exporting countries. After the wealth-acquisition phase, the wealth-preservation phase is the logical continuation of present policies. This second phase will prove to be much more difficult than the first one.

For one thing, time works against the oil-rich countries, just as it works against all wealthy countries, including the United States. The older generation in this country well remember the deprivations of the Great Depression in the early 1930s or during the Second World War. The memory of past poverty can never be completely erased, and this builds up defenses in consumption patterns that seem ridiculous to the younger generation. The fact is, young Americans, born into an affluent nation, don't know how to be frugal. This is only one of the great problems facing America in these days of diminishing national wealth.

Chances are this problem will be much more severe in the currently oil-rich nations, not because of a difference in their reaction to wealth, but because of the all-or-nothing nature of some of these nations' resource bases. Countries whose economies are built on a single resource are, to use a banking term, *overexposed* in a given commodity and, for that reason, are vulnerable.

By vulnerability it is not suggested that OPEC may break up under its own weight. Surely, the economists who have pursued that line of reasoning must have come to admit their error. The classical cartel vulnerability, preached by the economic profession throughout the West, has been shown to be no more than a classical error.[7] In the current context, vulnerability is defined as a kind of long-run exposure to the consequences either of the development of substitute energy or of a natural decline in Middle Eastern oil reserves. Either event is at least one, probably two, generations away. But one thing is absolutely sure: One of these events is bound to occur.

Expansion of OPEC Resource Bases

The first order of business in preserving current oil wealth is to investigate the possibilities of converting an oil-exporting nation from a single to a multiple resource base. This may be possible in some Middle Eastern countries, in others it may not. Where possible, it is the safest investment that can possibly be made, since it falls under the exclusive national, legal, and economic control of the resource-owning nation. Neither devaluations, expropriations nor inflations can take a direct toll out of such a resource.

On the other hand, one of the most wasteful errors a country can commit is to use its present wealth to build up an industry that cannot support itself in the long run. The classical example of this type of error in economic thinking is the developing country bent on having its own steel industry. Developing countries should have a steel industry *only* if that industry can sustain a viable and competitive economic life. The comparative advantage, discredited as it is today by Marxist charges of being a tool of capitalist exploitation, is nevertheless a very real concept that can only be ignored at great costs to those who do so.

Leading Middle Eastern contenders for development of alternative industries are Iran and Saudi Arabia. These two countries have vast excess funds to invest elsewhere. Thus, for them and other Middle Eastern oil producers, the question is where to invest and in what.

Real Wealth vs. "Paper" Investment

The first and most important point in this connection relates to an error in economic thinking that surfaces again and again in the mass of articles written by private and government experts on both sides.[8] The point is that the oil-exporting countries are not investing money, but *real* wealth, using money as a vehicle. This distinction is of the utmost importance. Without it, the investment problem is a banker's problem; with it, it becomes an economist's problem.

It will be pointed out later that investment in the so-called money markets, commonly (and, from the economist's point of view, erroneously) defined as the market of currency and all sight and short-term credit instruments (12 months or less), is at best a temporary vehicle—fraught with danger—that might be used by the oil-exporting countries while gearing up to develop their long-term investment capabilities. The long-term solution to the preservation of real wealth is real investment, not paper investment. The term "paper" applies to all currencies, including the U.S. dollar and Special Drawing Rights, as well as government bonds and nonconvertible corporate bonds.

The combined uncommitted oil revenues that will be generated in the Saudi Arabian Peninsula and in Iran will amount to about $43.4 billion per year. As was mentioned earlier, 80% of this amount, or $34.7 billion per year, could become available for investments. The last thing in the world the oil-exporting countries will do is to accept currency in payment. The Arab oil producers tried that once and lost $1.3 billion in the deal, due to the 1971 and 1973 dollar devaluations.[9] That was back in the days of low-priced energy, when the U.S. dollar was the universal standard of value.

A 20% dollar devaluation today, applied to only one year's uncommitted reserves, would cost Saudi Arabia alone more than $3.4 billion. The cost to Kuwait would be nearly $1 billion; Iraq and Abu Dhabi would each lose about $600-$700 million. Iran's losses would come out at close to $2.5 billion. Surely everybody, including the U.S. government, must understand that the oil-importing countries will have to do better this time around than to tender money straight from the printing presses.

The preceding figures emphasize the importance for oil-exporting countries to achieve high turn-over rates on their currency holdings. To reiterate, a 20% dollar devaluation applied to one year's uncommitted reserves of Saudi Arabia will cost that country $3.4 billion. If Saudi Arabia could achieve a portfolio position with no more than six months of uncommitted reserves in currencies, the cost would be $1.7 billion. A three-months' exposure would reduce that cost to $850 million. Given the necessary priorities and a year or two of operating experience, Saudi Arabia, as well as other oil-exporting countries, should be able to reach a three-month cash target.

In any event, there ought to be a cash target in the policy formulations of these countries, and that target ought to be communicated to oil-importing nations. This would facilitate long-term monetary planning on both sides and help to avoid disruptive crises.

Devaluations are not the only risks of amassing foreign currencies. Another problem is the less spectacular but equally destructive erosion of purchasing power of foreign currencies through inflation.

For example, U.S. inflation was about 12% per year in the Fall of 1974, and to this day, the worldwide inflation rate remains at the two-digit level in most nations. This kind of erosion, applied to Saudi Arabia's uncommitted yearly reserves, costs that country $2.0 billion every year. This is about three-fourths of Saudi Arabia's total oil revenues of 1972.[10]

This emphasizes even more dramatically the need to hold cash balances at an absolute minimum because the inflation loss on accumulated cash holdings cannot be recouped. If, for example, Saudi Arabia could hold its cash requirements to three months of uncommitted reserves, the inflation cost would be reduced by 75%, or $1.5 billion, to about $500 million per year.

Indexing the price of oil will protect the oil-exporting countries only against inflation losses on current or future transactions. It will provide no protection against past sales, where the proper protection is speed of investment and financial prudence.

For a non-U.S. owner to massively hold idle U.S. dollar balances in the United States in the form of money would be the poorest of all investments, since money (defined here as checking accounts and currency in circulation) yields no interest. As an alternative to money, credit may be used as a vehicle of investment. This may be an acceptable short-run alternative, but since it shares many of the flaws of money, it is no more than that. By credit is meant all interest-bearing instruments of all maturities, from sight to 20 or 30 years. Included are government bonds, corporate bonds and bank bonds, indexed or non-indexed, with or without a repurchase clause. Under this definition the Eurodollar market, for example, would be a credit market, and from the economist's point of view, quite properly so.

The U.S. government has been busy trying to convince oil-exporting countries that they should switch to U.S. government bonds if they don't like currency. And indeed many of these countries, unable to quickly invest elsewhere, are accumulating bonds.

To the extent that the investment portfolios of most OPEC nations will always contain some U.S. government bonds, there is no reason why the governments of oil-exporting nations should not press for negotiations aimed at protecting their U.S. government bond holdings through indexation. Indexed bonds admittedly do not presently exist in the United States, but they do exist in Europe, and there is no reason why such indexation could not be introduced here, at least on special bond-issues tendered to foreign governments.

Unless government bonds are indexed, they are about as poor an investment as currency. Like currency, they are vulnerable to both inflation and devaluation, but to the extent that a substantial portion of these bonds are nonmarketable, they cannot be dumped at times of stress, and that makes these bonds *more* vulnerable than currency. Of course, government bonds yield a return and U.S. checking accounts do not, so there is an incentive for bonds. Yet bonds, especially government bonds, currently yield a negative true rate of return, since their interest rates are below inflation rates.

There is another, more fundamental problem associated with currency and bonds: If held abroad, U.S. dollars are a liability to the U.S. economy, as are bonds. The only difference is that one yields a return and the other does not. Both are made available to the foreign investor by printing numbers on pieces of paper. They are, in effect, extremely cheap substitutes for real wealth. Because oil-exporting countries are concerned with preserving their real wealth, they cannot afford to accept paper wealth. They must eventually and predominantly invest in real foreign property—corporate or noncorporate equity, capital and land.

Equity Investment

Equity capital is not as vulnerable to inflation as are bonds, and that is its main attractiveness. What's more, appropriate monitoring of international money markets and balance-of-payments positions throughout the world, coupled with prudent and selective equity investment strategies, will keep international monetary crises at a minimum. This will do much to prevent unexpected devaluations, that other risk—besides inflation—that currency and bond holdings are exposed to.

Equity holdings exhibit characteristics all their own, especially if held by foreigners. The greatest risk here is nationalization or de facto nationalization, i.e., the freezing of repatriated profits. The latter policy, less overt but just as effective, leaves foreign capital nominally in the hands of the owner. That makes it less objectionable in world opinion.

Interestingly enough, the oil-exporting countries have much to learn in managing the equity capital they hold abroad. The first and most important consideration is that industrial nations did not get where they are now by playing games or by being idealistic promoters of wealth throughout the world. They got there by shrewd economic analysis, by hard work, by a willingness to take risks, and by a strict adherence to rules of the market. In short, they got where they are now by being capitalists.

The oil-exporting nations have now joined the ranks of the capitalists, defined here as "owners of capital," and not in its later and distorted ideological sense. The only way to preserve that capital is by playing the game by capitalistic rules. Indeed, most oil-producing nations are doing just that, from necessity, not from ideological conviction, except that the terminology used is often very much a-capitalistic, even anticapitalistic, sometimes downright Marxist. Still, nationalization, if applied against you, is objectionable, and it is a risk that the oil-exporting countries have to learn to guard against.

There are obviously nationalization-prone countries in the world, some expressly, others implicitly so. And then there are, just as obviously, havens from nationalization. Between these two poles, with countries like Peru on the one side and Germany or the United States on the other, lies a continuum of governmental attitudes that has to be researched, defined, and analyzed.

Clearly, capital will preferably flow to places where it has enjoyed traditional safety: Europe, North America, Japan (although that's a hard country to penetrate) and Brazil, to name a few. But even among the safe countries, there are many differences making for better or worse investments.

The internal inflation rate, for example, in addition to being a wealth-destroyer, is an indicator of a government's economic competence. Labor problems, to give another example, can positively wreck an economy, as they are doing in the United Kingdom. This alone, coupled with heavy nationalizations of domestic capital, pretty well marks Great Britain as one of the poorest places to invest among any of the Western European countries. Yet, it has been and it remains a primary target for Middle Eastern investment dollars, perhaps because of historical hangups.

Among the most difficult problems relating to international investments, and especially to nonliquid equity investments, are those where domestic legislation conflicts with the foreign investor's legislation. For example, a French subsidiary of a U.S. firm once received a sizable order for machinery to be exported to Mainland China. That was at a time when the United States and China did not maintain trade relations. Thus, delivery on that order was clearly contrary to U.S. law. But it was not contrary to French law, and many French jobs were at stake at the time.

The case went through the courts, French courts of course, and the company was ordered to start production. The American executive involved in that case showed a great deal of diplomacy, and the problem was eventually resolved to everybody's satisfaction—with machinery being delivered by a U.S. firm to a then declared enemy of the United States. This problem may easily arise for an Arab subsidiary located in the United States and involving the sale of equipment to, say, Israel.

Beyond the ideological conflict of interest, there are others. For example, in 1974 the U.S. Congress overruled a presidential veto and thereby provided that railroad workers employed by private U.S. concerns be given retirement pay out of U.S. tax revenues. If Arab capital were heavily invested in the U.S. trucking industry, it would probably consider the railroad subsidy unfair. Yet there would be little the foreign owners could do about it. Indeed, the danger exists that those industries that are in foreign hands might be consistently denied subsidies for their U.S. workers, while similar subsidies are given to other industries. This kind of discrimination is best prevented by spreading investments over many industries.

Many more examples of discriminatory U.S. policies could be cited, such as the subsidizing of strikes in the United States by making strikers eligible for the food stamp program. On this issue, OPEC investors and U.S. capitalists may come to experience a mutuality of interests, and they may form a common front against U.S. labor unions.

Discriminatory bias of this type has also been experienced by many U.S. companies operating abroad. For example, in many instances these companies have been ordered to submit to union rules that did not apply elsewhere, or to grant fringe benefits that were not available to resident workers outside the U.S. firms. When these tactics are applied, the foreign corporation can do nothing else but accept them. Again, it would behoove a foreign investor to know in advance the countries where such policies are likely to be implemented. Countries that grossly abuse this power ought to be avoided.

Wise investment requires the acquisition and analysis of good information. On that score, no country offers better opportunities than the United States, through its Securities and Exchange Commission (SEC). Periodic reports required by the SEC are available to the public and generally are more detailed accounts than the usual annual reports to the shareholders. For example, the SEC 10-K report contains financial and corporate information on an annual basis. The 10-Q report serves a similar purpose, except that it is issued and available quarterly. Both reports contain unscheduled material events on corporate changes deemed of importance to the SEC. These reports and other information are available for a small reproduction fee. Computerized ratio analyses, disclosure publications, and a wealth of other information can be used to quickly spot a likely candidate for acquisition, in any desired industry and in any desired geographical location within the United States. As always, availability of data is in direct proportion to a country's economic development, a fact that tends to mitigate against investment in developing countries.

It's a fact of life that Americans are getting uneasy about the prospects of large-scale OPEC equity investments in the United States. Multinational corporations with many joint ventures or 100%-owned subsidiaries all over the globe are fearful of becoming OPEC subsidiaries. For years the multinationals have failed to understand why foreign individuals or governments should not welcome them because, as they saw it, their injection of capital and technological know-how could not do anything but help host countries. Yet, faced with the similar prospects of OPEC investment in the United States, they are anything but elated. They are, in fact, downright worried, and because they are worried, they are setting up defenses against takeovers.

On the other hand, OPEC members are as short on financial managers as they are long on petrodollars. To take over and to run a large U.S. corporation requires managerial expertise specifically geared to the legal and political U.S. environment. A particular problem area, previously mentioned and wholly new to OPEC managers, is labor relations.

At first, OPEC countries may avail themselves of the services of the trust departments of large banks to handle their investments. In fact, some of the countries are doing this now.[12] Eventually, however, as soon as specific U.S. expertise has been developed among OPEC managers, this arrangement will probably give way to the establishment of OPEC-owned and administered U.S.-based investment and management corporations. To back up a crash training program of OPEC managers, the establishment of an Arab University in the United States (as well as one in Europe and on other continents), and limited to M.B.A.-type programs of the highest quality, ought to be considered. In a way, this would be the reverse of the American Universities located outside the United States.

As mentioned previously, U.S. corporations are anything but anxious to be acquired by OPEC interests. Their defenses are manifold, including lobbying for foreign equity restrictions on corporate shares. Motivated by fears of domestic and foreign takeovers, small U.S. corporations are increasingly going private, where they are much less vulnerable.

Large corporations contract the services of specialists who analyze, on a day-to-day and hour-to-hour basis, stock market transactions on their corporate shares and who report instantly any unusual moves. These alarm signals then trigger carefully prearranged defense strategies including instant telephone calls to major shareholders, personal letters from corporate presidents to the remaining shareholders, delaying court actions on technical violations of SEC regulations, and other measures.[13]

As time goes by, U.S. corporations will become more efficient in the fight for independent corporate existence. But there is always one convincing counterargument: money. Given a sufficient incentive, shareholders will part with their corporate shares, but these will become more expensive.

There is one thing that the American public, and certainly the American government, *must* understand, and it only requires a look at the oil-wealth problem from the oil exporter's point of view: Given the vulnerability of currencies and bonds, equity holdings are the only acceptable long-term answer to the current transfer of wealth to oil-exporting nations. If the equity route is denied to the oil exporters, and there are noises in Congress that it might be, then they have absolutely no alternative but to cut back oil production. This will provoke angry reactions, to be sure, in oil-importing countries who will fail to see, or prefer not to admit, that it was their own policies that drove the exporters to these extreme steps.

Money has been said to be a veil. Underlying money is barter, and from the oil exporter's point of view, that barter spells out "oil for machinery and plants and all sorts of productive equipment." It does not spell out, as it has to some extent in the past, "oil for paper." It is to be hoped, for the sake of both the oil exporters and importers, that the money veil does not blind the oil-importing nations to this fundamental truth.

iii. **OPEC Petrodollars and U.S. Stagflation**

As we have seen, the disadvantage of investing in U.S. corporate or government bonds is that these do not protect against inflation or devaluation, and they are part of a deeply troubled credit market. These troubles originate mostly with ill-advised money and credit policies using the rate of interest as an indicator of the tightness of money and credit. That is poor economic theory and it makes for poor economic policy, since the rate of interest, in addition to measuring the scarcity of credit, also measures the inflation rate. For example, an 8% loan will yield $8 on an investment of $100, in the case of zero inflation. However, if the inflation rate rises to 7%, that same investment will have to yield $15.56, so that the invested capital plus interest will not be eroded through inflation. Thus, inflation brings about rising interest rates and the pressure is on to do something, and do it quickly. Since everybody knows that an increase in the supply of money and credit will reduce the rate of interest, the pressure is on the U.S. Federal Reserve System to ease its monetary constraints. This will be done, eventually, and the interest rate will come down, but the increased money supply will bring about a long-term increase in price levels and, therefore, still higher interest rates which, again, call for easy-money policies. The vicious cycle is complete.

In addition to raising interest rates through the inflation rate, the U.S. government puts direct upward pressure on the rate of interest by massive credit demands to finance its own budgetary deficit. For example, the $65.6 billion deficit in fiscal 1976 corresponded to a net addition to domestic credit demands of nearly $260 million every working day. When the government enters the credit market with these demands, it finds itself competing with corporations and households, and the ensuing competition will drive up interest rates, thereby creating the illusion of tight money and credit and triggering demands for further easing of monetary policies. This will drive up prices and embark the U.S. on the vicious inflationary cycle once more.

Of course, there is a rather extremely reliable measure of the scarcity of money: the price level itself. It seems absurd to pretend that money is scarce when the inflation rate is high, yet this claim is made all the time by highly respected professionals. The value of money can only be eroded when there is too much of it in relation to the goods and services in the economy. We all knew this to be self-evident, until the economists explained it to us differently.

From past observation, and on the basis of an estimation of what the future holds, why worry about the OPEC nations destabilizing U.S. financial markets when we already have a government with a remarkable record of doing this? What's more, with a $50-plus billion deficit annually, the U.S. government is doing it considerably more massively than OPEC's $11 billion (see section *i* of this chapter). Indeed, a good case can be made for allowing these $11 billion to purchase U.S. equities, for this will release an equivalent amount of U.S. funds to the credit market, thereby preventing the interest rate from rising as high as it would otherwise.

OPEC investment in the United States is not the menace it is made out to be. After all, having a vital stake in this economy, OPEC countries will be just as eager to see it perform satisfactorily as we are.

References

1. *The Wall Street Journal*, Sept. 19, 1974, p. 3.
2. Merklein, H.A., "The energy crisis: Some causes and effects," *World Oil*, April 1974, pp. 91-94.
3. *Petroleum Intelligence Weekly*, Jan. 7, 1974, p. 9.
4. *Middle East Economic Survey*, Vol. XVII, No. 46, Sept. 6, 1974. p. 4.
5. "Stockholders' Equities from the Fortune 500 List," *Fortune Magazine*, May 1974, pp. 230-257.
6. Chase Manhattan Bank, *The Petroleum Situation*, Jan. 30, 1976.
7. Merklein, H.A., "The energy crisis: Some causes and effects; Part 3: A look at international oil, the OPEC cartel and windfall profits," *World Oil*, March 1974, pp. 47-72.
8. For one example, out of many, see P.T. Leach, "U.S. urged to stimulate flow of petrodollars," *The Journal of Commerce*, Oct. 16, 1974, p. 1.
9. Merklein, H.A. "The energy crisis . . . " *World Oil*, op. cit., p. 42, and G. Corm, "Arab capital funds and monetary speculation," *Arab Oil and Gas*, Aug. 16, 1973, pp. 24-30.
10. "Some statistics from SAMA Reports for 1973," *Middle East Economic Survey*, Vol. XVII, No. 52, p. (i).
11. "Money market gets big influx of petrodollars," *The Wall Street Journal*, Oct. 18, 1974, p. 21.
12. Demaree, A.T., "Arab wealth, as seen through Arab eyes," *Fortune*, April 1974, pp. 108-190.
13. "Company executives shore up defenses against take-overs," *The Wall Street Journal*, Oct. 21, 1974, p. 1.

6
Energy and Indexation

i. How Exporters Can Preserve the Purchasing Power of Crude Oil

Ultimately, the price of any commodity, including oil, is determined by the interplay of demand and supply. Taken as a group, the OPEC members currently control the supply, while the oil-importing countries represent the demand side.

Different oil-importing countries have different options in their respective demands for oil. The United States, with its large potential oil and gas reserves, in addition to huge coal deposits and well-developed nuclear technology, has alternative energy options that are simply not available to most other oil-importing nations. Yet all of these potential sources of energy are useless, unless a determined effort is made to develop them.

The United States has paid a great deal of lip service to developing new oil and gas reserves, but there has been no positive action. For several years after the oil embargo, gas prices continued to be controlled and old oil, relabeled lower-tier oil, sold at 35% of world prices while the U.S. Congress was debating horizontal and vertical divestiture of the oil industry.

Congress has followed up Project Independence with what may be called Project Incompetence, and while it did a great deal of talking and some energy legislating, almost all of it counterproductive, imports as a percent of total oil consumption have risen in spite of tripled world crude oil prices.

On the supply side, there exists an administered price of crude oil, at this writing set at $11.51 per barrel of the benchmark Arabian light crude but almost certain to go up before this book is published. There is every indication that this price or the new price can remain in force indefinitely, perhaps with the help of a slight reduction in crude oil deliveries. In fact, considerably higher prices can be attained, as the 1973-1974 embargo has amply demonstrated.

The question discussed in this chapter is this: If the OPEC members wish to preserve the purchasing power of their oil, what must they take into consideration and what techniques are best suited to meet such an objective?

The issue at hand is wealth preservation. For example, in July 1974, Kuwait was producing close to 2 million bpd, mostly for export. That production represented a certain amount of wealth that could be used to purchase goods for imports to Kuwait. In July 1975, Kuwait's oil was still selling at the 1974 price. Because oil prices are generally expressed in U.S. dollars, this means that a given export rate of Kuwaiti oil produced the same dollar payments in mid-1974 and 1975. Yet these same U.S. dollars did not buy equivalent goods for imports because the goods produced in oil-importing countries and offered for sale to OPEC nations had experienced substantial price increases through inflation.

The price of crude did not inflate from mid-1974 to mid-1975, while the prices of most other goods did. That is why the purchasing power of oil, in terms of other goods, declined in that period. And that is why the OPEC nations suffered a relative wealth decline.

However, inflation is not the only source of disruption of the purchasing power of a given commodity such as oil. Since many commodities, and especially crude oil, are traded internationally, the rates of exchange of the trading countries' currencies are also of great importance. For example, at an assumed government take of $10 a barrel, 100,000 barrels will produce $1 million U.S. If the dollars are spent exclusively in the United States, the exchange rate problem is not present, since both the sale of oil and purchases of import goods are denominated in U.S. dollars.

Suppose, however, the money is used to purchase German-made goods. In that case, the inflation rate in Germany becomes a critical determinant of the purchasing power of oil. But since oil sales are U.S. dollar denominated and German exports are (generally) denominated in Deutsche marks, the dollar-Deutsche mark exchange rate (absent in the pure U.S. dollar case) also becomes an important factor, and, in particular, the change of that exchange rate with time.

For example, the sum of $1 million U.S., generated through the sale of 100,000 barrels of crude oil represented 2.555 million Deutsche marks in mid-1974 (end of June, to be precise). By mid-1975, that same $1 million U.S., produced by that same sale of 100,000 barrels of oil, represented 2.355 million Deutsche marks, DM200,000 less than a year earlier, simply because the DM-$U.S. exchange rate had declined from 2.555 to 2.355.

Because both the inflation rates in oil-importing countries and (with the exception of the United States) the variation over time of international exchange rates have the potential of eroding the purchasing power of crude oil, a protecting mechanism must be devised against each of these forces.

In a *floating exchange rate* system the international strength or weakness of a country's currency is affected by its domestic strength or weakness. Inflation at home creates deteriorating exchange rates abroad, provided a country's domestic inflation rate exceeds that of its international trading partners. An example will illustrate this point.

Suppose Country A deals with Country B. Let A's annual inflation rate be 15%, compared to 5% for B. In this case A's prices, including prices of export goods, exhibit an annual increase over B's prices by approximately 10%. If trade between Countries A and B starts at equilibrium in a given base year, say 1970, A's exports to B earn exactly the amount of B-dollars to finance A's imports from B, and vice versa. The two countries have an equilibrium trade balance; no trade surplus and no deficit exists. At the initial A/B exchange rate of, say 2:1, this condition will *not* remain unchanged.

Given the 10% inflation differential, the year of 1971 will find A's goods approximately 10% more expensive (expressed in A-dollars) than B's goods (expressed in B-dollars). At the fixed 2:1 exchange rate, the citizens of B now find that they can still purchase one A-dollar for 2 B-dollars as before, but that the A-dollars so purchased buy 10% less than they used to. Actually, they buy 15% less, but the purchasing power of the B-dollars used to purchase A-dollars has eroded by 5%, so that the net effect is 10%.

In short, the citizens of Country B find imports from Country A about 10% more expensive in the market place. As a result they will buy fewer imports, i.e., Country A's exports to B will decline. A's resulting balance-of-trade deficit will put downward pressure on the international value of that country's currency.

Similarly, B's exports to Country A become cheaper to A-citizens who, as a result, will buy more goods made in B. This aggravates the currency drain from Country A and the downward pressure on A-dollars is intensified. In the end, caught between declining exports and rising imports, A's currency will float down, thereby reestablishing a new equilibrium exchange rate, say 1.95:1.

However, this is not the end of the story. Unless remedial policy action is undertaken to curb the relatively high inflation rate in Country A, the downward pressure on A's currency will persist. The exchange rate might be 1.88:1 in 1972, and 1.82:1 in 1973, and so on.

The point to remember is that oil prices must be established in such a way as to prevent wealth erosion of OPEC nations through inflation or through variations in exchange rates, but that the latter are affected by the former.

This section and the next will deal only with the inflation question; the exchange rate issue will be discussed in section *iii*. Suffice it to say briefly that the exchange rate issue stems from the question of what actual or artificial unit of account to use in billing oil sales. In particular, many OPEC members have proposed to use the Special Drawing Right as a unit of account. The pros and cons of such a procedure also will be discussed in section *iii*, as will the confusion surrounding the use of a national currency either as a unit of account or as a store of value.

Protection against differential inflation is best accomplished through an indexing procedure. Because rising prices are generally involved on both sides, both must be taken into consideration. Moreover, since the purchasing power of a given export commodity (or of a group of commodities) is to be protected in terms of what it will import, only exports and imports and their price variations are to be considered, not overall inflation rates.

OPEC members presently export, almost exclusively, petroleum and petroleum products. This group of commodities is best represented by crude oil prices. The issue at hand, then, is how to price crude oil with time so that a given barrel of oil will purchase the same amount of imports needed by OPEC nations. The indexing mechanism to use in this case is based on the so-called terms of trade.

The terms of trade of one country with another country, or of one commodity group relative to another commodity group, is nothing but the ratio of their weighted average prices, expressed in real terms. What matters here is not so much the price itself, but the *movement* of that price over time. This is best explained with a hypothetical example involving the well-known country of Slobovia (located 500 miles east of the Equator) and the United States, its exclusive trading partner. Suppose Slobovia exports coffee, other agricultural goods and finished goods in accordance with Table 6-1, which also shows its imports from the United States both in 1970 and 1976.

As can be seen in Table 6-1, Slobovia's market basket of export goods contains 100,000 units of coffee, 50,000 units of agricultural goods, and 200,000 units of finished goods at respective unit prices of U.S. dollars 2,000, 3,000 and 1,500. Their total exports generate hard currency in the amount of $650 million U.S., which exactly finances the importation of 50,000 units of capital goods at a unit price of $4,000 U.S.; plus 100,000 cars costing $2,000 U.S. each and 250,000 other types of goods selling at $1,000 U.S. each, columns (1) through (4).

Columns (5) through (7) show the change of events in 1976. Since the basic question, always, is whether the real purchasing power of a given commodity basket has changed and by how much, the composition of that basket must be left unchanged, as in column (5). This isolates price changes as the sole cause in changing terms of trade.

Table 6-1
Slobovia's Terms of Trade

(1)	Year 1970			Year 1976		
	Volume (x10³)	Unit Prices (x10³)	Total Exports (x10⁶)	Volume (x10³)	Unit Prices (x10³)	Total Exports (x10⁶)
(1)	(2)	(3)	(4)	(5)	(6)	(7)
Exports						
Coffee	100	$2.00	$200	100	$2.00	$200
Other agric....	50	3.00	150	50	3.00	150
Fin. goods	200	1.50	300	200	2.00	400
			$650			$750
Imports						
Cap. goods ...	50	$4.00	$200	50	$5.00	$250
Cars	100	2.00	200	100	3.00	300
Other	250	1.00	250	250	2.00	500
			$650			$1,050

Column (6) shows what happened to world prices in 1976. For example, world prices of coffee and other agricultural goods have remained the same, while the price of Slobovia's finished export goods has risen from $1,500 to $2,000 per unit. Thus, the same export basket that brought Slobovia $650 million U.S. in 1970 will now bring $750 million U.S. in 1976. So far so good; Slobovia appears to have made gains. Or has it?

To answer that question, it becomes necessary to determine what that hard currency now buys in world markets. In terms of Slobovia's 1970 import basket, if that is less, Slobovia will have suffered, regardless of the dollar amounts involved. If it is more, it will have gained. The technical phrase is that Slobovia's terms of trade have deteriorated in the first case and improved in the second. As it turns out, Slobovia's export basket is inadequate in 1976 to finance the import basket it was capable of financing in 1970. An analysis of the price movements shows why. Of Slobovia's three export commodities, only one (finished goods) rose in price, and that one by 33%, while prices of all commodities Slobovia is importing rose, some substantially—capital goods went up by 25%, cars by 50%, and other goods doubled in price.

In fact, taking all price changes into consideration, Slobovia's export basket can purchase only 750 ÷ 1,050, or 71.4% in 1976 of what it used to purchase in 1970. Thus, the purchasing power or terms of trade of Slobovia's export basket has deteriorated by 1 ÷ .714, or 40%.

To offset this loss, Slobovia would have to increase the 1976 prices of its export commodities by 40.0%, if it can do this, given the world demand and supply situation. Thus, Slobovia's prices would have to be as shown in Table 6-2. Because the technique illustrated in Table 6-2 involves an adjustment of current prices, this is called *Current Period Indexation,* as opposed to *Base Period Indexation,* which will be taken up in section *ii*. As can be seen in Table 6-2, increasing the 1976 unit price of coffee and of finished goods from $2,000 to $2,800 and the unit price of other agricultural goods from $3,000 to $4,200 (i.e., by 40%) will assure stability of the purchasing power of Slobovia's export basket. Thus the wealth-preservation objective is accomplished by indexing on the basis of a country's terms of trade.

In theory, then, and ignoring the world demand and supply situation, wealth preservation through indexing presents no particular problem. The question now arises *how* such a procedure might be implemented in practice. If at all possible, the practical application of indexation should use existing price series. Fortunately for OPEC, such series are available, as we shall see next.

ii. Practical Aspects of Crude Oil Price Indexation

We have just seen that Slobovia's 1976 export prices would have to be adjusted to keep its terms of trade from slipping. Since the purchasing power of Slobovia's export goods had declined by 40.0%, the method calls for a price increase of these goods by precisely that percentage. Column (9) in Table 6-2

Table 6-2
Slobovia's Terms of Trade: Current Period Indexation

| (1) | Year 1970 | | | Year 1976 | | | | |
	Volume (x10³)	Unit Prices (x10³)	Total Exports (x10⁶)	Volume (x10³)	Unit Prices (x10³)	Total Exports (x10⁶)	Adjusted Prices (x10³)	Adjusted Exports (x10⁶)
	(2)	(3)	(4)	(5)	(6)	(7)	(8)	(9)
Exports								
Coffee	100	$2.00	$200	100	$2.00	$200	$2.80	$280
Other agric...	50	3.00	150	50	3.00	150	4.20	210
Fin. goods ...	200	1.50	300	200	2.00	400	2.80	560
			$650			$750		$1,050
Imports								
Cap. goods ..	50	$4.00	$200	50	$5.00	$250		
Cars	100	2.00	200	100	3.00	300		
Other	250	1.00	250	250	2.00	500		
			$650			$1,050		

reflects the restored purchasing power of Slobovia's export basket in relation to its imports.

In practice, it is often more convenient to work with given base periods. Rather than adjusting prices up and down from current levels, it is generally easier to refer price adjustments back to some preestablished base period. This is particularly important if the use of existing indices is involved, since all index series follow the base period pattern.

In terms of Tables 6-1 and 6-2, this means that Slobovia's wealth-preservation objective can be achieved by using an appropriate upward adjustment of its 1970 prices of export goods, rather than using the relevant 1976 prices.

Here is the logic. Table 6-1 shows that the total hard currency demands of Slobovia for the purpose of financing given import volumes rose from $650 million to $1,050 million, or by 61.5%, in the period 1970-1976, while the country's ability to generate hard currency through exports rose only from $650 million to $750 million, or by 15.4%. An increase of Slobovia's export prices by 61.5%, relative to 1970, would boost the value of export goods by just the right amount, so that the country's export basket will once again finance the purchase of its import goods as it did in 1970. This is illustrated in Table 6-3.

Table 6-3
Slobovia's Terms of Trade: Base Period Indexation

	Year 1970			Year 1976				
	Volume (x10³)	Unit Prices (x10³)	Total Exports (x10⁶)	Volume (x10³)	Unit Prices (x10³)	Total Exports (x10⁶)	Adjusted Prices (x10³)	Adjusted Exports (x10⁶)
(1)	(2)	(3)	(4)	(5)	(6)	(7)	(8)	(9)
Exports								
Coffee	100	$2.00	$200	100	$2.00	$200	$3.23	$323
Other agric...	50	3.00	150	50	3.00	150	4.85	243
Fin. goods ...	200	1.50	300	200	2.00	400	2.42	484
			$650			$750		$1,050
Export price index			100			115.4		161.5
Imports								
Cap. goods ..	50	$4.00	$200	50	$5.00	$250	—	—
Cars	100	2.00	200	100	3.00	300	—	—
Other	250	1.00	250	250	2.00	500	—	—
			$650			$1,050		
Import price index			100			161.5		

The overall result of current period indexation (Table 6-2) and base period indexation (Table 6-3) is the same: both methods reestablish the purchasing power of export goods to 1970 levels. The only difference is that individual prices of export goods are adjusted to different levels. This has definite economic implications, since adjusted prices in Table 6-2 more nearly correspond to current world market conditions than adjusted prices of Table 6-3. In other words, domestic and international market distortions are minimized by using the method of current period indexation.

For countries specializing in the export of only one commodity group, such as the OPEC members, this distinction is academic. OPEC nations export essentially only one commodity: petroleum and related products. Petroleum and its products thus would replace Slobovia's three commodity groups, with the result that individual price differences cannot emerge, since only one individual price is listed. Thus, current period indexation and base period indexation are identical for OPEC members. This, and the convenience of using available price series, makes the base period method the better one to use in practice.

Also shown in Table 6-3 are export and import price indices. These indices are 100 in the 1970 base year. The 1976 *export price index* was calculated as follows:

$$I_{exp.} = \frac{750}{650} \times 100 = 115.4$$

As mentioned earlier, this means that export prices have risen over the relevant period by 15.4%. Similarly, the 1976 *import price index* is

$$I_{imp.} = \frac{1,050}{650} \times 100 = 161.5$$

This denotes a 61.5% increase in import prices.

OPEC members obtain most of their imports from industrialized countries. Thus, if OPEC members could index their export prices on the basis of imports from industrialized countries, they could for all practical purposes achieve terms-of-trade stability, thereby maintaining the purchasing power of crude and related products, and checking the wealth decline now in progress.

There exists a series that corresponds very nearly to OPEC import price movements. It is published by the International Monetary Fund's *International Financial Statistics* under the heading "Changes in Export Prices—Industrial Countries" (see page 32 of the October 1975 issue of that publication, for example).

On the premise that, on average, changes in export prices from industrial countries are a reasonable reflection of changes in import prices of OPEC nations (with the notable exception of military hardware, which is not reflected in the IMF index), this particular series can be used with acceptable accuracy as an indexation base for world crude oil prices. Table 6-4 shows how crude oil prices would have moved, had this indexation mechanism been used after the embargo.

The indexed crude oil prices in column (5) of Table 6-4 are illustrated in Figure 6-1. The solid line in that figure indicates the actual price of the Light Arabian marker oil in 1974-1976. The steep rise midway through the curve reflects the 10% increase of October 1, 1975, which pushed the price from $10.46 to $11.51 per barrel.

The dashed curve shows what the price of crude oil would have been had OPEC introduced an indexation plan on the basis of the cited IMF series. It is immediately apparent from Figure 6-1 that indexation is a two-sided sword: it cuts both ways. For example, export prices on goods originating in industrial countries showed very steep increases in the first year after the embargo, rising from 170 in the second quarter of 1974 to 197 in the first quarter of 1975. Accordingly, the indexed crude oil price would have risen sharply, to a level of $12.12 per barrel in early 1975. A temporary reversal in price trends during the last three quarters of 1975 would have pushed the price of crude oil to a low of $11.38 per barrel in the fourth quarter of 1975. At that point, with the 10% price increase in effect, the actual price of crude oil was slightly above the equivalent indexed price. Indications are that export prices are rising again, though not as rapidly as in 1974, so that the indexed crude oil prices in the first and second quarters of 1976 show a moderate increase.

Table 6-4
How Indexation Would Have Affected Post-Embargo Oil Prices

Year	Quarter	Export Price Index	Crude Oil Prices, $/bbl.		Foregone Price Increment $/bbl.
			Actual	Indexed	
(1)	(2)	(3)	(4)	(5)	(6)
1974	II	170	10.46	10.46	—
	III	176	10.46	10.83	0.37
	IV	184	10.46	11.32	0.86
1975	I	197	10.46	12.12	1.66
	II	196	10.46	12.06	1.60
	III	187	10.46	11.51	1.05
	IV	185	11.51	11.38	−0.13
1976	I	188	11.51	11.57	0.06
	II	189	11.51	11.63	0.12

Source: IMF, *International Financial Statistics,* various issues.

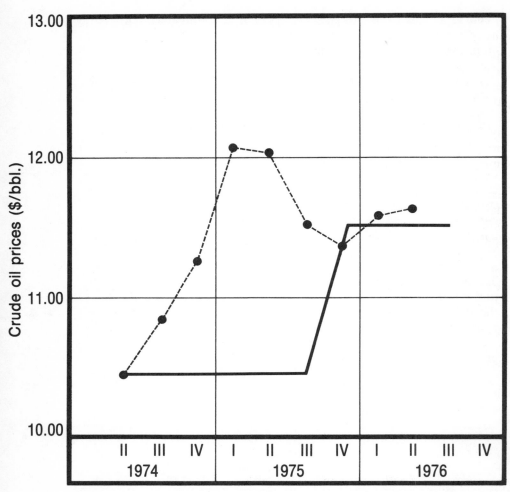

Figure 6-1. *Indexed and non-indexed oil prices, 1974-1976 (U.S. dollars per barrel).*

To be noted also is the lag problem. As a general rule, the more highly developed a country, the faster it is capable of assembling and disseminating internal economic data. Thus, the industrial country export index offers the advantage of minimum lags. The countries included in this category are the United States, Canada, Japan, Austria, Belgium, Denmark, France, Germany, Italy, The Netherlands, Norway, Sweden, Switzerland, and the United Kingdom, a total of 14 countries.

Still, the April 1976 issue of the *International Financial Statistics* shows the fourth quarter 1975 export price index on the basis of eight country indices available at that time. Missing are six countries, some of them relatively high inflators: Canada, Austria, France, Italy, Norway, and Sweden. If, after perhaps a total lag of six months, the missing country indices become available, the resulting overall export index is likely to be higher.

Column (6) in Table 6-4 reflects the per-barrel loss in wealth experienced by oil-exporting countries in 1974-1975 due to their failure to use an indexing procedure. Expressed in terms of total losses (Table 6-5), these figures are impressive.

Both the dollar volumes and the growth of loss increases are startling. For example, the largest OPEC producer, Saudi Arabia, lost $3.14 million per day in the first quarter, $7.29 million per day in the second quarter, and $11.3 million per day in the third and fourth quarters following the base period. Qatar, the smallest producer listed here, experienced losses of $190,000 per day, $450,000 per day, and $700,000 per day, respectively, over the same period. Total Saudi losses in the third and fourth quarters of 1974 and the first and second quarters of 1975 amounted to $3.005 billion. Losses of other OPEC members are listed in Table 6-6.

The numbers in Table 6-6 speak for themselves. Note that these are terms-of-trade losses. They do, however, include losses or gains resulting from exchange rate fluctuations. This will be explained in more detail in the next section.

No wonder OPEC nations are concerned about indexation. Had indexation been introduced in mid-1974, the losses shown in Table 6-6 would have been recovered.

As a practical matter, a workable indexation plan will use only one index. That index will necessarily be a weighted average. Therefore, it will benefit some OPEC members a little more than others. But the important point is that it will bring substantial benefits to all. For a good discussion of individual country indices, see the study entitled "Changes in OPEC Import Price Indices," Economic Research Institute of the Middle East in Tokyo, *Middle East Economic Survey*, Nov. 14, 1975, Supplement.

Deviations from the average index are experienced for at least two reasons: First, the use of an average price increase of goods exported from industrial countries implies that all OPEC nations import from industrialized countries only, and that their percentage imports from the various industrial countries are the same. This is, of course, not the case. The second reason relates to the fact that many countries and therefore many currencies are involved, while the average terms-of-trade index is based on U.S. dollars. More about this later.

The one prohibitive problem with individual country indices is that they will eventually lead to substantial price differentials for crude oil, thereby set-

Table 6-5
OPEC Incomes Lost Through Terms-of-Trade Deterioration

Country	Third quarter 1974[a] Production 10⁶ bpd†	Income Losses $10⁶/day	Income Losses Total ($10⁶)	Fourth Quarter 1974[a] Production 10⁶ bpd†	Income Losses $10⁶/day	Income Losses Total ($10⁶)
Mideastern OPEC members						
Saudi Arabia	8.48	3.14	289	8.48	7.29	671
Iran	6.02	2.23	205	6.02	5.18	477
Kuwait	2.55	0.94	86	2.55	2.19	201
Iraq	1.89	0.70	64	1.89	1.63	150
U.A.E.	1.69	0.63	58	1.69	1.45	133
Qatar	0.52	0.19	17	0.52	0.45	41
Other major OPEC members						
Venezuela	2.98	1.10	101	2.98	2.56	236
Nigeria	2.26	0.84	77	2.26	1.94	178
Indonesia	1.38	0.51	47	1.38	1.19	109
Libya	1.52	0.56	52	1.52	1.31	121
Algeria	1.01	0.37	34	1.01	0.87	80

Country	First Quarter 1975[b,c] Production 10⁶ bpd†	$10⁶/day	Total ($10⁶)	Second Quarter 1975[b] Production 10⁶ bpd†	$10⁶/day	Total ($10⁶)
Mideastern OPEC members						
Saudi Arabia	7.08	11.3	1017	7.08	11.3	1028
Iran	5.35	8.6	774	5.35	8.6	783
Kuwait	2.08	3.3	297	2.08	3.3	300
Iraq	2.22	3.6	324	2.22	3.6	328
U.A.E.	1.69	2.7	243	1.69	2.7	246
Qatar	0.44	0.7	63	0.44	0.7	64
Other major OPEC members						
Venezuela	2.35	3.8	342	2.35	3.8	346
Nigeria	1.79	2.9	261	1.79	2.9	264
Indonesia	1.31	2.1	189	1.31	2.1	191
Libya	1.51	2.4	216	1.51	2.4	218
Algeria	0.96	1.5	135	0.96	1.5	137

[a] Average 1974 production data.
[b] Average 1975 production data.
[c] Reflects index of 196, prior to adjustment to 197 in October 1976.
†Production data include domestic consumption, causing income losses to be slightly overstated.

Table 6-6
Total Losses Through Terms-of-Trade Declines
(Third and Fourth Quarters 1974—First and Second Quarters 1975)

Mideastern OPEC Members	$ Millions	Other OPEC Members	$ Million
Saudi Arabia.............	$3,005	Venezuela	$1,025
Iran......................	2,239	Nigeria	780
Kuwait...................	884	Indonesia	536
Iraq......................	866	Libya	607
U.A.E...................	680	Algeria	386
Qatar	185		
Total	$7,859		$3,334
Total OPEC losses: $11.9 billion			

ting the stage for divisive arbitrage operations or, at best, for squeezing high-index countries out of world oil markets. Whether individual country indices are used or not, the increase in nominal crude prices that will result from any kind of terms-of-trade indexation ultimately calls for restriction in output by OPEC nations. Since that is in the long-run interest of OPEC members, such a restriction should be feasible. After all, oil production in many OPEC nations has peaked out, and the instincts of self-preservation will be sharpened by declining oil exports. Member countries will bring increasing pressure on Saudi Arabia to cooperate in a voluntary program of production allocation, thereby making indexation a workable reality.

In view of the relatively inelastic demand for oil imports, fostered among other things by the inability of the U.S. government to implement meaningful incentive programs at home, such an allocation program will be enormously profitable for OPEC members and for non-OPEC oil exporters.

iii. Indexation and Currency Devaluations

The International Monetary Fund series on export indices, mentioned earlier, is based on prices expressed in U.S. dollars. Thus, the dollar price of a given export good from, say, Germany may rise for one of two reasons:

1. At fixed DM/$ exchange rates the good may become more expensive in terms of Deutsche marks. This is the inflation case that was discussed in the preceding two sections.

2. Export prices may remain fixed in terms of Deutsche marks, but the U.S. dollar may be devalued (or the DM upvalued in relation to the U.S. dollar). This is the exchange rate case to be discussed.

Suppose OPEC is dealing with two hypothetical countries, Upper and Lower Slobovia. Their currencies, the Upper Slobofranc (USF) and the Lower Slobofranc (LSF), both exchanged for U.S. dollars at the rate of 5:1 in 1965, the base year. Moreover, both countries exhibited price *increases* in their export goods by approximately 2% per year. However, the USF was *upvalued* twice in relation to the U.S. dollar: On January 1, 1969, by 10%, and on January 1, 1973, by 25%. The LSF, on the other hand, was *devalued* twice, relative to the U.S. dollar: the first time on January 1, 1968, by 15%, and the second time on January 1, 1973, by an additional 5%. Table 6-7 shows the impact on the export price indices of both countries.

The upper section of Table 6-7 represents the hard-currency case of Upper Slobovia, whose export price index, USF base, rises by about 2% per year, column (2). This is illustrated by the solid line in Figure 6-2.

The export price index in column (2) may be thought of as the price of a representative market basket of export goods. In 1965, that basket sold for USF 100; in 1966, the same market basket sold for USF 102, and so on, until its price rose to USF 120 in 1975.

At the original 5:1 exchange rate, the market basket price of USF 100 corresponds to $20 U.S. Given the 2% export price increase in terms of the USF and the fixed USF/$ exchange rate, the *dollar price* of Upper Slobovia's export basket also rose initially by 2% per year. However, in 1969 the USF was upvalued by 10% (or the $ U.S. devalued).

Subsequent to the USF upvaluation, the 1969 price of Upper Slobovia's export basket stood at USF 108, in accordance with the 2% export price increase. But since the U.S. dollar was devalued, it now took *more* U.S. dollars to buy one USF's worth of export goods. Given the new exchange rate of 4.5:1, the equivalent dollar price of Upper Slobovia's export basket jumped to $24 U.S. A similar upward move in the U.S. dollar price of Upper Slobovia's export basket was again experienced in 1973, when the dollar was devalued a second time, column (4).

Column (5) in Table 6-7 (upper section) is a repetition of the equivalent dollar prices of Upper Slobovia's export baskets, expressed in terms of an index and using 1965 as the base year. This is Upper Slobovia's export price index, based on U.S. dollars. This index is illustrated in Figure 6-2 by the broken line which shows clearly the substantial increases resulting from the 1969 and 1973 upvaluations of the Upper Slobofranc.

The relevance of this U.S. dollar-based export price index now comes into focus. If OPEC had negotiated indexed crude oil sales payable in Upper Slobofrancs, column (2) in Table 6-7 or the solid line in Figure 6-2, would represent the appropriate index. If OPEC had negotiated an indexed sale payable in U.S. dollars, column (5) in Table 6-7 or the broken line in Figure 6-2 would represent the relevant export price index. To be noted is the fact that the dollar-base index was calculated from column (2), i.e., taking into ac-

Table 6-7
Export Price Indices—Upper and Lower Slobovia

	Upper Slobovia—USF Upvaluations ($ Devaluation)			
Year	Export Price Index, Based on USF	Exchange Rate USF/$	Equivalent Dollar Prices[a]	Export Price Index, Based on U.S. $
(1)	(2)	(3)	(4)	(5)
1965	100	5.00	20.00	100
1966	102	5.00	20.40	102
1967	104	5.00	20.80	104
1968	106	5.00	21.20	106
1969	**108**	**4.50** [b]	**24.00**	**120**
1970	110	4.50	24.40	122
1971	112	4.50	24.90	124
1972	114	4.50	25.30	126
1973	**116**	**3.38** [c]	**34.30**	**172**
1974	118	3.38	34.90	175
1975	120	3.38	35.50	178

	Lower Slobovia—LSF Devaluations ($ Upvaluation)			
Year	Export Price Index, Based on LSF	Exchange Rate LSF/$	Equivalent Dollar Prices[a]	Export Price Index, Based on U.S. $
(1)	(2)	(3)	(4)	(5)
1965	100	5.00	20.00	100
1966	102	5.00	20.40	102
1967	104	5.00	20.80	104
1968	**106**	**5.75** [d]	**18.40**	**92**
1969	108	5.75	18.80	94
1970	110	5.75	19.10	96
1971	112	5.75	19.50	98
1972	114	5.75	19.80	100
1973	**116**	**6.04** [e]	**19.20**	**96**
1974	118	6.04	19.50	98
1975	120	6.04	19.90	100

[a] Rounded to nearest 10¢.
[b] 10% upvaluation of Upper Slobofranc.
[c] 25% upvaluation of Upper Slobofranc.
[d] 15% devaluation of Lower Slobofranc.
[e] 5% devaluation of Lower Slobofranc.

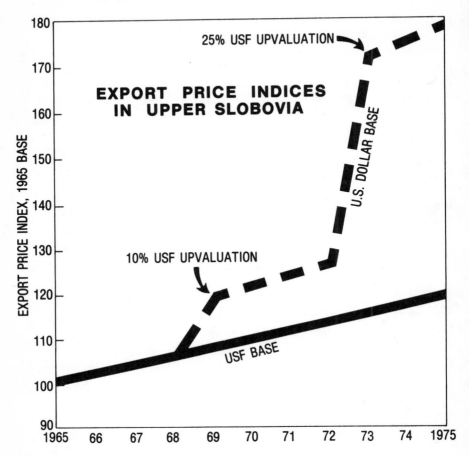

Figure 6-2. *Export price indices in Upper Slobovia.*

count the USF denominated 2% annual price increase. Thus, the dollar index reflects both export price movements and exchange rate fluctuations.

Suppose for the sake of argument that a barrel of oil cost $5 (USF 25) in 1965. By 1968, that same barrel will command the same purchasing power in Upper Slobovia as it did in 1965, if it is sold for $5 x 1.06 = $5.30 or USF 25 x 1.06 = USF 26.50.

After the USF upvaluation in 1969 (the U.S. dollar devaluation) the two indices diverge, and for good reason. If the contract calls for payment in Upper Slobofrancs, a 2% increase in the USF price of crude oil will bring OPEC the additional purchasing power that is needed to overcome Upper Slobovia's export price increase. If the contract calls for payment in U.S. dollars, a 2% price increase would not be enough, since it takes 10% more dollars to purchase Upper Slobofrancs, and 2% more of these francs are required to compensate for Upper Slobovia's increase in export prices. Thus, the 1969 indexed price in Slobofrancs will be USF 25 x 1.08 = USF 27, while the 1969 dollar price is $5 x 1.20 = $6. These and other indexed prices are shown in Table 6-8.

The situation in Lower Slobovia is the exact opposite of Upper Slobovia. OPEC is dealing with a *soft* currency that may or may not be accepted in world monetary markets. If the Lower Slobofranc is not acceptable in world markets, the U.S. dollar may be both a better medium of exchange and a better unit of account. If used as a medium of exchange, collection is made in actual U.S. dollars, which can be used in countries other than Lower Slobovia for international trade purposes. If used as a unit of account, especially in a multi-currency system as we have it in the real world, the resulting crude oil price will tend to fluctuate less violently than otherwise.

The lower section of Table 6-7 shows Lower Slobovia's export price index both on the basis of LSF's (column 2) and on the basis of U.S. dollars (column 5). These indices are plotted in Figure 6-3. To be noted is the 2% increase in each index, as long as the LSF/$ exchange rate is at parity, that is, until 1967. The 1968 devaluation of the Lower Slobofranc has no bearing on the LSF index, which continues to rise at about 2% annually (solid line).

Table 6-8
Indexed Crude Oil Prices—Upper Slobovia

Year	Export Price Index		Crude Oil Prices	
	USF Base	$ Base	USF/bbl.	$/bbl.
(1)	(2)	(3)	(4)	(5)
1965	100	100	25.00	5.00
1966	102	102	25.50	5.10
1967	104	104	26.00	5.20
1968	106	106	26.50	5.30
1969	**108**	**120**	**27.00**	**6.00**
1970	110	122	27.50	6.10
1971	112	124	28.00	6.20
1972	114	126	28.50	6.30
1973	**116**	**172**	**29.00**	**8.60**
1974	118	175	29.50	8.75
1975	120	178	30.00	8.90

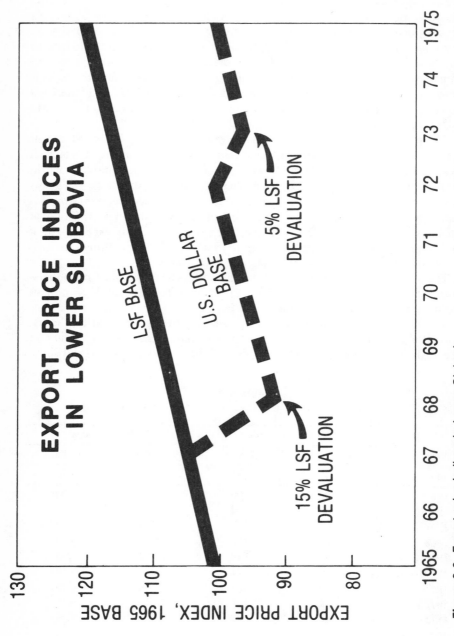

EXPORT PRICE INDICES IN LOWER SLOBOVIA

LSF BASE

U.S. DOLLAR BASE

5% LSF DEVALUATION

15% LSF DEVALUATION

EXPORT PRICE INDEX, 1965 BASE

130 — 120 — 110 — 100 — 90 — 80

1965 66 67 68 69 70 71 72 73 74 1975

Figure 6-3. Export price indices in Lower Slobovia.

However, the dollar-base index declines in response to the LSF devaluation (broken line), since it now takes fewer U.S. dollars to purchase devalued Lower Slobofrancs for purposes of financing OPEC's imports from Lower Slobovia. The indexed crude oil prices for the Lower Slobovia case are listed in Table 6-9.

The individual country export price indices listed in the monthly volumes of the *International Financial Statistics* (International Monetary Fund) are calculated in accordance with the method used in developing the U.S. dollar based export price indices for Upper and Lower Slobovia, column (5) in Table 6-7. This is best illustrated with an actual example, say France, for the period 1967 to 1974. The calculation follows the Slobovia pattern, and the results are shown in Table 6-10.

Columns (5) and (6) in Table 6-10 reflect essentially identical dollar-based export price indices for France. Thus, the calculation procedure outlined in this section is the one used by the International Monetary Fund and the resulting individual country indices do in fact account for both inflationary price movements and exchange rate fluctuations. The suggested export price index for industrial countries is simply a weighted average (using the 1970 value of exports as weights) of the individual indices. As was mentioned in the previous section, the overall export price index is based on a total of 14 industrial countries.

Certainly, indexation is a workable wealth preservative. What's more, once indexation has become an established policy, with automatic quarterly adjustments, the ensuing increases in the price of crude oil can no longer be

Table 6-9
Indexed Crude Oil Prices—Lower Slobovia

Year	Export Price Index		Crude Oil Prices	
	LSF Base	$ Base	LSF/bbl.	$/bbl.
(1)	(2)	(3)	(4)	(5)
1965	100	100	25.00	5.00
1966	102	102	25.50	5.10
1967	104	104	26.00	5.20
1968	**106**	**92**	**26.50**	**4.60**
1969	108	94	27.00	4.70
1970	110	96	27.50	4.80
1971	112	98	28.00	4.90
1972	114	100	28.50	5.00
1973	**116**	**96**	**29.00**	**4.80**
1974	118	98	29.50	4.90
1975	120	100	30.00	5.00

Table 6-10
Export Price Index for France—1963 Base

Year	Export Price Index, Based on FF	Exchange Rate FF/U.S. $	Equivalent Dollar Prices	Export Price Index U.S. Dollar Base	
				Calculated	Listed in IFS
(1)	(2)	(3)	(4)	(5)	(6)
1963	100.0[a]	4.937[b]	$20.26	100.0[c]	100.0[d]
1967	106.9	4.937	21.65	106.9	107.0
1968	106.0	4.937	21.47	106.0	106.0
1969	113.9	5.143*	22.15	109.3	110.0
1970	125.7	5.554	22.63	111.7	111.5
1971	133.0	5.554†	23.95	118.2	118.3
1972	134.4	5.045	26.64	131.5	131.3
1973	147.8	4.439**	33.30	164.4	165.0
1974	184.8	4.810	38.42	189.6	190.5

[a] From IFS Country Tables, Line 74.
[b] Used Trade Conversion Factor 1963—1973, IFS Country Tables, Line ra. Used Par Rate/ Market Rate for 1974-IFS Country Tables, Line rf.
[c] Calculations based on method outlined in this section.
[d] Listed in Export Price Tables—France, IFS.
*FF devaluation, third quarter 1969.
†U.S. dollar devaluation, last quarter 1971.
**Second U.S. dollar devaluation, first quarter 1973.

blamed on OPEC. They simply reflect a built-in response to the inability of industrial countries to contain price increases on export goods.

As things are now, each time OPEC members meet to discuss ways of protecting themselves from world inflation, they are under tremendous pressures—economic and implied military pressures—to resist increases in the price of crude oil. Yet, this is the *only* means of self-protection. The simple solution is to agree, once and forever, on a self-adjusting policy of price indexation; then prices will no longer have to be set—they follow from prevailing world market conditions.

There is a precedent to this type of pressure evasion—in the United States. This is the case of the floating prime rate which certainly opens interesting possibilities concerning a floating price of crude oil. It is the case of major U.S. commercial banks coming under heavy fire from the Administration and from Congress for occasionally raising their prime interest rates on loans to large business establishments.

It did the bankers no good to point out that they were only responding to market pressures. The fact is that, pressures or no pressures, they did determine what the prime rate should be at given points in time, and they set the rate accordingly. Of course, they could not set the prime rate at arbitrary levels, but that was an argument neither the Administration nor the Congress cared to pay attention to. In short, a scapegoat was needed, and the bankers did just fine in that capacity.

Then the bankers found a way to redirect the blame for rising interest rates where it belonged: with the U.S. government. Beginning in November 1971, the large banks announced that *they* would no longer set the prime interest rate. They published a formula by which the prime rate would set itself, based on several credit market rates, notably interest rates on government bonds. From that point on, every time the interest rate on government bonds rose in the United States, so did the prime rate—automatically. And there was no one left to blame but the government itself.

The commercial bankers simply took the position that it was easy enough to get the prime rate down: Reduce the interest rate on government bonds, and the prime rate will follow—*automatically*. The relevance of this case to "floating crude oil prices" is both obvious and interesting.

The topic that has been ignored until now is the question concerning use of an acceptable unit of account in selling OPEC oil. To be sure, the IMF export price index is based on prices expressed in U.S. dollars. This and the fact that the U.S. dollar has been notoriously overpriced in the past, leading to the devaluations of 1971 and 1973, which cost Arab OPEC members alone some $1.3 billion, gave rise to a number of questions and objections against the use of a dollar-based index. Actually, fears in that regard are unfounded, since the U.S. dollar only serves as a numéraire, that is, as a unit of account. That does not make the dollar a medium of exchange, and actual payments may well be made in terms of other hard currencies. Indeed, OPEC nations would be well advised to make it a point to extend their billings over all hard currencies. Such a procedure will reduce the predominance of the U.S. dollar in world markets and thus prevent a recurrence of the quasi-monopolistic hold the U.S. dollar once had in international monetary markets, especially in the period prior to 1973, when the international monetary system was adhering to fixed exchange rates.

There is a way, of course, to get away from the dollar base altogether—by using the SDR as a unit of account. This would not greatly change the resulting index, since the SDR itself is based on 16 currencies, 13 of which are also represented in the currency basket of the 14 industrial nations, and the remaining three represent relatively small countries whose combined weight in the SDR valuation is only 4.0%. A list of nations represented in the industrial and SDR indices and of their relative weights is given Table 6-11.

Whether the crude oil price index is SDR-based or U.S. dollar-based is of no particular importance, since the two bases yield essentially the same result. For example, an economic environment characterized by 2% annual price increases of export goods and a 15% devaluation of the French Franc yields the following indices, relative to the preceding year's index of 104.0:

Table 6-11
IMF "Industrial" and "SDR" Nations

Nation	Weight in Dollar Index	Weight in SDR
United States	20.8%	33.0%
Germany	16.4	12.5
United Kingdom	9.3	9.0
Japan	9.3	7.5
France	8.7	7.5
Canada	8.0	6.0
Italy	6.3	6.0
Netherlands	5.7	4.5
Belgium	5.6	3.5
Sweden	3.3	2.5
Switzerland	2.4	—
Denmark	1.6	1.5
Austria	1.4	1.0
Norway	1.2	1.5
Australia	—	1.5
Spain	—	1.5
South Africa	—	1.0
	100.0	100.0

SDR-based 104.9
U.S. dollar-based 104.8

Detailed calculations leading to these indices are published in the *Proceedings of the Third International Conference on U.S. and World Energy Sources,* October 18-19, 1976, Boulder, Colorado, and need not be repeated here.

And now a final point. It has been argued by some rather eminent economists that OPEC is not a cartel, since its members do not restrict output. The no-cartel argument lacks analytical rigor, and this becomes more and more apparent in the market place, where the issue of output restriction must sooner or later be confronted by OPEC.

OPEC has no need to be on the defensive, particularly since untenable defensive arguments cloud the issues and create confusion inside OPEC as well as outside. After all, there are many other cartels, notably the governments of oil-importing countries that used to tax away nearly 50% of the total revenue generated by the production, shipment, refining and sale of one barrel of crude oil—all this in return for zero services.

The best of all cartels was the old Krupp cartel, especially prior to World War II. That cartel produced goods (guns and related military equipment) that were sold to opposing armies who put these goods to use destroying each other. Krupp did rather well under this arrangement.

More seriously and to the point, the very survival of OPEC as a unit depends on output restriction. What's more, restricting output is bound to be a profitable undertaking, since the demand for oil by importing countries is relatively (not completely) inelastic. The technical expression is that the elasticity of demand for oil is less than unity. The meaning of this term is, simply, that a given percentage increase in the price of crude oil will induce a smaller percentage cutback in oil imports.

For example, and the figures used here are purely illustrative, a 10% increase in prices might induce no more than a 5% cutback in oil imports. That makes price increases a profitable venture, provided OPEC can agree on how to allocate sales cutbacks among member countries. If each OPEC member, for example, would agree to cut back its exports by 5%, it could benefit from a 10% price increase, for a net increase in sales revenue of approximately 5% (4.5% to be exact). That is a very tempting situation and it is difficult to see, given the exhaustible nature of their resources, how the OPEC nations can resist it very long.

7
Energy
Alternatives

i. Alternate Fossil Fuels

A global evaluation of fossil fuel alternatives indicates none can contribute markedly to the U.S. energy supply in the near future:

—Secondary and tertiary recovery are somewhat overrated in terms of recoverable reserves. However, they do constitute our most immediate sources of oil.

— Oil shale reserves are high, but enormous capacity problems exist. This source is not likely to be a substantial contributor to U.S. energy.

— Tar sands are unimportant to the United States as a whole.

— Synthetic oil and gas from coal may be the best organic fuel sources in the United States, but they are subject to substantial capacity problems and development lags. Coal and its derivatives will be first among alternative fuels until the perfection of nuclear power.

It is apparent that future U.S. production of crude oil will not come from a single source but will be derived from all possible sources. The size and quality of oil reservoirs or deposits that can be tapped in the future will depend on the economic environment, process development, research, and many other factors. Political interference via price roll-backs or price controls in a generally inflationary environment can greatly impede the development of any or all alternative energy sources.

In 1970, about 22% of total U.S. oil consumption came from foreign sources. In 1973, the percentage had risen to 35, then declined to 30% because of the embargo in the first half of 1974. In early 1976, U.S. imports were up again, at 42% of U.S. consumption. Thus the growth of U.S. dependence on foreign suppliers had come to exceed even the most pessimistic expectations.

The Arab oil embargo set the U.S. energy industry scrambling for new resources. There are various energy alternatives available, all costly, many of them of limited and short-term value. To the extent that the data are known, this section deals with the following energy alternatives:

1. New developments in secondary and tertiary recovery methods from oil and gas reservoirs
2. Oil shales
3. Tar sands
4. Synthetic oil and gas from coal

A word first on the term "reserves." This stock variable has been used, and abused, in the press and on Capitol Hill, largely because there are several kinds of oil reserves, each with its own definition. "Proved" reserves of crude oil are defined by the American Petroleum Institute (API) as follows:[1]

> Reservoirs are considered proved if economic producibility is supported by either actual production or conclusive formation tests. The area of an oil reservoir considered proved includes: (1) that portion delineated by drilling and defined by gas-oil or oil-water contacts, if any; and (2) the immediately adjoining portions not yet drilled but which can be reasonably judged as economically productive on the basis of available geological and engineering data. In the absence of information on fluid contacts, the lowest known structural occurrence of hydrocarbons controls the lower proved limit of the reservoir.

It is clear that proved reserves are measured with a good deal of conservatism. Indeed, a more descriptive term of "proved" reserves would be "bankable" reserves, for these are the reserves that oil companies carry on their books and which they can use as collateral for loans. For example, oil companies sometimes have their "proved" reserves calculated annually by consulting firms, precisely because their reserves are used as collateral on corporate indebtedness, a situation that calls for a third-party appraisal.

Use of a conservative accounting standard by the petroleum industry is wholly justifiable. Furthermore, the practice is as old as the industry itself. Only if the petroleum industry had recently switched its reserve accounting standards could it be fairly accused of withholding reserve information or of slanting it in such a way as to create the illusion of a shortage. Of all the charges that have been voiced against the oil industry, the charge of conspiratorial changes in reserve accounting procedures has never been mentioned, simply because there is no basis for such an allegation.

After proved reserves, the next larger category is "probable" reserves. In addition to "proved" reserves, "probable" reserves include reserve estimates in a given geological structure in areas beyond well locations immediately adjoining producing wells. Oil people are reasonably sure that they will find oil or gas in the probable areas, but they literally can't bank on it.

"Potential" reserves are vaguest of all. They are reserves that are *thought* to exist in certain regions, without proof or even firm indications. For example, the continental shelf off the Atlantic Coast is believed to have "potential" reserves, even though there is currently no production. In this discussion, and unless otherwise stated, reserves will be held to mean proved reserves.

Like inventories, reserves are a stock variable. Taken by themselves, oil or gas reserves mean little, as do inventory levels. It is common business practice to express inventories in more meaningful terms, either as inventory turnover rates or in terms of so many days' or months' worth of inventories. The latter measure, called reserve/production ratio or reserve-life index, expressed in so many years, is generally used in the oil industry.

What matters here are changes in the reserve-life index, and these changes, invariably, have been in a downward direction. Given U.S. oil reserves of some 35 billion barrels[2] and a U.S. consumption rate of approximately 17.5 million barrels per day,[3] the U.S. reserve-life index now stands at 5.5 years. This means that we would run out of oil in 5.5 years if:

1. There were no oil imports, and
2. There were no way to switch into other forms of energy, and
3. There were no further additions to existing reserves through U.S. oil exploration, and
4. All oil wells could be produced at capacity until exhausted, i.e., ignoring the natural decline of a well's oil production with time.

None of these conditions hold, and the United States is not going to consume its last drop of oil 5.5 years from now. Still, as little as 10 years ago, the reserve-life index stood at 7.8 years. Its rapid decline in recent years is a matter of grave concern.

As will be seen later, neither the reserve-life index nor the quantity of proved reserves are of much help when it comes to unorthodox oil sources such as oil shales, tar sands, and coal. Both the reserves and the reserve-life index may be large in these categories, but there are definite limits on the rate of exploitation. Moreover, since each of the sources requires extensive mining operations and construction of a petrochemical plant the size of an average refinery, more or less, they are characterized by considerable production lags.

As a nation, the United States has committed itself, through the hastily conceived Project Independence, to do without oil or gas imports by 1980. This leaves only two energy sources: nonconventional forms of U.S. energy and increased U.S. oil and gas reserves. Where, precisely, does the United States stand on these?

Secondary and Tertiary Recovery

Contrary to popular belief, neither secondary nor tertiary recovery methods make much of a dent in the current energy crisis. Nor do these methods have the potential, if fully exploited, to relieve the shortage.

Generally, and exceptions are few and far between, only oil fields lend themselves to secondary or tertiary recovery methods. Gas fields are already subject to high recovery efficiencies (70-80%), and thus do not offer a great deal of unrecovered gas in the reservoir to work on. What's more, low pressure, not low productivity, terminates the life of a gas well.

If gas prices warrant installation of scrubbing devices and compressors prior to discharge either into gas plants or pipelines, the bottomhole pressure of gas reservoirs can be reduced and recovery increased by primary production. The one exception is a very tight gas reservoir having a permeability of a few millidarcies. In these fields, where primary recovery is marginal, secondary and tertiary recovery are usually out of the question.

That leaves oil reservoirs as the only hydrocarbon deposits subject to secondary or tertiary recovery methods. Almost invariably, secondary recovery is held to mean waterflooding. There are too many variants of this process to list here. However, reserves in carbonaceous rock matrices do not ordinarily lend themselves to waterflooding as efficiently as do sandstone matrices. That only leaves consolidated or unconsolidated sands, or about 70% of all past and present U.S. oil reservoirs. In terms of oil originally in place, this corresponds to some 306 billion barrels potentially subject to secondary recovery.[4]

In addition to these geo-technical difficulties there are legal problems with waterflooding (as well as with tertiary recovery methods). In some states, the legal structure does not encourage or even permit unitization of a field, an indispensible prerequisite to waterflooding.

Texas, to give one example, does not allow mandatory unitization. Many a barrel of oil that could be recovered through unitization and subsequent waterflooding remains unrecovered. The major oil companies have always pushed for mandatory unitization, i.e., for more efficient recovery methods, but they were consistently thwarted by small independents or individual lease holders. After the embargo, the same majors were charged with having engineered the energy crisis by withholding production.

Suppose there were no legal problems and all known sand or sandstone reservoirs, whether still in production or not, were converted or reactivated to encompass waterflooding. What would be the increase in recoverable reserves?

On the sweeping assumption that, on average, primary recovery through the artificial lift stage is about 18%, secondary recovery averages 35%, and using 75% of total oil originally in place in all non-carbonaceous U.S. reservoirs, from first commerical production through 1971, or some 75% of 306 billion barrels, the increase in total recoverable reserves as a result of secondary recovery comes out to be 39 billion barrels. Adding this quantity to current primary reserve estimates of 35 billion barrels yields a total of 74 billion barrels. At present consumption rates, this would raise the reserve-life index from 5.5 to 11.5 years. So, an increase in the reserve-life index by six years is all that can be expected from secondary recovery—and that only through considerable efforts and expenses and at a slow rate.

Broadly speaking, tertiary recovery is what is left by way of recovery methods after natural depletion and waterflooding. The four predominant recovery methods are steam injection, miscible slugs, chemical floods, and in-situ combustion.

Steam injection, in particular the so-called huff-and-puff method of intermittent injection in a producer, has resulted in considerable improvements in production in some local applications, notably in California. But for the nation as a whole, it offers no hope for increasing recoverable reserves.

High hopes were once associated with the miscible slug method and with chemical floods. All of the many variants of the miscible slug method (presolvent water injection, cosolvent water injection, gas-water injection following solvent placement, to name a few) suffer from one basic drawback: the solvent, invariably, is a hydrocarbon product. As such, it rises in price as the price of crude oil rises, and at least proportionately so.

Thus, an increase in crude oil prices, which would ordinarily stimulate use of secondary and tertiary recovery methods, has no such effect on the miscible slug method. That method has been found too expensive in the past, at least insofar as full-scale adoption in the oil industry is concerned, owing to the enormous volumes of hydrocarbons that need to be injected.

That same method will be found to be too expensive in the future. In fact, many oil company research departments have all but given up work in this field, and miscible slugs will be ignored here because of their highly questionable value as a future recovery booster.[5]

That leaves in-situ combustion, which received a great deal of attention throughout the '60s, only to become less well thought of in the early '70s. Rising crude oil prices have again focused attention on in-situ combustion.

In situ combustion involves injection of oxygen, or, more commonly, of air, into an oil reservoir, where the presence of oxygen and hydrocarbons permits ignition. The effectiveness of producing oil from strategically located producers rests on two phenomena: first, the oil to be produced is heated and becomes less viscous; second, burned-off gases, mostly CO, CO_2, steam, as well as light-end-hydrocarbons, form a sort of in-situ miscible slug with attendant high sweep efficiencies, displacing in-place oil and pushing it toward producing wells.

As with waterfloods, fissures and channels in carbonaceous reservoirs cause early breakthroughs and unacceptably low sweep efficiencies. Thus, only sands and sandstones lend themselves to in-situ combustion, and not all of these. The rock matrix must exhibit desirable depth-permeability combinations to assure the technical and economic success of in-situ combustion.[6]

If half the U.S. oil reserves originally contained in sands and sandstones, or about 153 billion barrels, lent themselves to in-situ combustion treatment, and if an average recovery efficiency of 50% of oil remaining after secondary recovery is accepted as reasonable, then additional recoverable reserves after primary and secondary recovery are given by the following equation:

$R = f_1 \ f_2 \ r_3 \ [1.00 - (r_1 + r_2)] \times O$, where
R = Total tertiary recovery potential
f_1 = Fraction of sands and sandstones, relative to all oil reservoirs (0.71)
f_2 = Fraction of noncarbonaceous oil reservoirs subject to potential in-situ combustion (0.5)
r_1 = Primary recovery efficiency (18%)
r_2 = Increase in recovery efficiency due to waterflooding (17%)
r_3 = Tertiary recovery efficiency, 50% of oil remaining in place after secondary recovery
O = Proved oil initially in place, billions of barrels (429)

Thus,
$R = 0.71 \times 0.5 \times 0.5 \ [1.00 - (0.18 + 0.17)] \times 429 = 49$ billion barrels

In-situ combustion, then, provides 25% more than the reserve additions potentially available through waterfloods. The reserve-life index, taking into consideration both waterflooding and in-situ combustion, would rise from the present 5.5 years to about 19 years. Decidedly, secondary and tertiary recovery methods, even if applied to previously discovered U.S. oil reserves, can at best buy time—all of 13.5 years.

Heavy crude oil reserves, not mentioned previously and for the most part not included in the 429 billion barrels of proved oil initially in place, are also likely candidates for tertiary recovery, especially in-situ combustion. These oils, with an API gravity of 25° or less, represent a recoverable reserve of around 20 billion barrels if only the most promising reservoirs are considered (class 1), with another 10 billion barrels in less promising reservoirs (class 2), using a 40% recovery efficiency and in-place figures as published by the U.S. Department of the Interior.[7]

Taken together, these class 1 and class 2 heavy-oil reserves would add a little less than five years to the U.S. reserve-life index, raising it from 5.5 years without to 24 years with secondary and tertiary recovery of all, including heavy, crude oils. The additions to U.S. oil reserves through secondary and tertiary recovery and through heavy oils are tabulated in Table 7-1.

Table 7-1
U.S. Reserve Additions

	Recoverable Reserves (billions of barrels)		Reserve-Life Index (years)	
	Per Category	Total	Per Category	Total
Primary Oil	35	35	5.5	5.5
Secondary Oil	39	74	6.0	11.5
Tertiary Oil	49	123	7.5	19.0
Heavy Oil	30	153	5.0	24.0

Shale Oil

Physical existence of known energy resources does not constitute reserves in the economic sense. That is a lesson the U.S. Interior Department learned over the last few decades in its role as administrator of federally held oil shales. In 1968, when Interior put up for bid three oil-shale leases in Colorado, two of these brought a combined offer of $500,000, and the third lease could not be given away. Interior had hoped to realize $80 million on the lease auction, and it turned down the lone offer as too low.

In January 1974, the U.S. energy picture had changed completely. Oil was being auctioned off abroad at unheard-of-prices of $17+ per barrel, and the uncontrolled portion of domestic oil was nearly $10 per barrel in the U.S. market. This provided the needed incentive to go after nonconventional sources of energy, among them shale oil.

On January 7, 1974, Interior held another auction on a 5,100-acre Colorado tract. Seven bids were received and the tract was awarded to Indiana Standard and Gulf Oil for $210.3 million.[8] At $41,319 per acre, that was the highest offer tendered to that time on any oil lease. The highest previous offer had been $36,808 per acre, a joint Exxon-Mobil-Champlin bid submitted two weeks earlier on a 5,760-acre tract off the Florida coast. Definitely, the oil industry was interested in oil, any oil, including shale oil.

Nevertheless, Interior, in view of the many unknowns surrounding the recovery of shale oil, planned to auction off no more than a total of six leases for the time being. Their plans called for close monitoring of these six pilot projects on federal lands, plus any that might be implemented on private lands, before further leases were to be auctioned. One area of intensive study, but not the only one, was the environmental effects of shale oil production.

U.S. oil shale formations are relatively well defined, both as to location and oil content. The only deposit presently of interest is the Eocene Green River formation in Colorado, Utah, and Wyoming, where it covers some

16,000 square miles (Figure 7-1). The oil content can go as high as 35 gallons per ton in prime areas.

The term "oil shale" is a misnomer. The rock matrix is not a shale but a marlstone, containing not oil but kerogen, an organic substance that yields shale oil on heating. Being solid, oil shale has to be mined and crushed prior to heating in a retort. In-situ retorting is a technical possibility, the economics of which are still very uncertain.

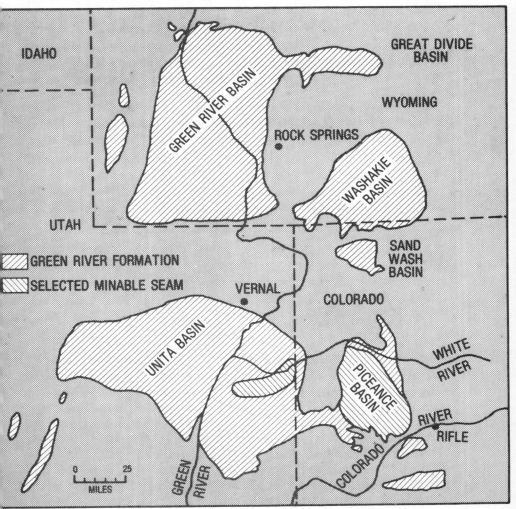

Figure 7-1. *Where U.S. shale oil reserves are concentrated. the deposit of most commercial interest is the eocene age green river formation which covers some 16,000 square miles. oil content is as much as 35 gallons per ton in prime areas.*

The Oil Shale Task Group of the National Petroleum Council has subdivided oil shales into four categories:

Class 1: Average yield of 35 gallons per ton over an unbroken interval of at least 30 feet.

Class 2: Average yield of 30 gallons per ton over an unbroken interval of at least 30 feet.

Class 3: Yield of 30 gallons per ton, but not found in a 30-foot continuous interval.

Class 4: Poor yields and less well-defined thicknesses.

Classes 1 and 2 constitute primary targets. Class 3 represents potentially recoverable reserves, if the energy crunch continues and provided the government allows development by letting the price of oil rise in response to market forces.

Table 7-2 shows shale oil reserves by states and classes. A 60% recovery coefficient was used in calculating reserves from in-place volumes, based on assumed subsurface mining operations.

Table 7-2
Shale Oil Reserves
(Billions of barrels)

States	Class 1	Class 2	Class 3	Total
Colorado	20	50	100	170
Utah	—	7	9	16
Wyoming	—	—	2	2
Total	20	57	111	188

Source: National Petroleum Council, *U.S. Energy Outlook,* Vol. II, p. 159.

If all shale were recovered by strip mining, at a 100% recovery coefficient, total shale oil reserves would grow to 314 billion barrels, or the equivalent of a 50-year autonomous U.S. oil supply. Between these two extremes, a reasonable estimate would allow for a 50-50 recovery by subsurface mining and strip mining. This assumption yields a recoverable reserve of about 252 billion barrels, enough to supply the United States at 1973 rates for 40 years.

However, a huge volume of underground oil-bearing shale is in the nature of the Heidelberg wine keg. Its volume notwithstanding, it is the size of its spout that determines the rate at which it can be drained. There are several barriers,—legal, environmental, technical, and economic—that must be overcome in chipping away the underground oil shale.

The legal barrier involves the fact that some 25% of Class 1 and 2 reserves have clouded titles. Most ownership disputes are between private prospectors and their heirs, and Interior. Until a 1920 revision of the Mining Law of 1872, prospectors could stake a claim on minerals they discovered on government lands. The Mineral Leasing Act of 1920 closed oil shale land to further private claims. Interior maintains that many pre-1920 claims were technically incorrect, some were fraudulent, and many more were forfeited for various reasons after they had been legally granted.

Be this as it may, 75% of Class 1 and 2 resources have undisputed ownership. Recoverable reserves underlying these lands are equal to 72 billion barrels of shale oil, about twice presently proved U.S. primary crude reserves and certainly enough to start mining operations in earnest.[9]

The environmental barrier revolves around the problem of solid residue disposal. The retorting process leaves a friable or dusty spent residue which, even after compaction, is larger in volume than the original oil shale from which it came, based on 30-gallon-per-ton shale.

A shale-oil output rate of one million bpd (about 6% of 1973 annual U.S. oil consumption) would create some 400 million cubic yards of residue, enough to bury Manhattan under 17 feet of waste material. To carry the Manhattan analogy further, if it had been possible to provide the total 1973 U.S. crude oil consumption from oil shales, the resulting residue would stack 300 feet at year's end, and three years later the Empire State Building would have disappeared.

The waste-disposal problem of shale oil production is formidable, even if half the residue (the upper limit) is returned to the mines. If surface mining is used, all waste material will eventually be returned to the pit, but in the intervening decades a temporary disposal problem persists.

Stacked on the surface, the residue presents a problem that goes beyond esthetics. Major dangers are dust storms, water contamination, and flash floods. The waste material is rich in soluble salts, thus posing the problem of both surface and groundwater contamination. The spent material is inimical to vegetation, and re-vegetation programs will probably be very expensive.

The very size of oil shale related mining operations represents a formidable technical barrier. To produce one million bpd, a total of 1.5 million tons of oil shale would have to be mined, crushed, and retorted daily. This amount approaches the total annual U.S. coal output. If all oil production had to be replaced by shale oil, the required combined oil shale mining operations would be 15 times larger than current U.S. coal mining, and concentrated in the States of Colorado, Utah and Wyoming.

The development lag between lease purchase and startup of oil shipments is at least four or five years, probably more. The five leases awarded in 1974 will not begin to relieve the U.S. energy pinch until 1980 if developed vigorously, which is doubtful, since crude oil prices are being held back. Even then, total production will be at most 100,000 to 150,000 bpd per lease, or 0.5 to 0.75 million bpd total. That is less than 4% of U.S. oil consumption in 1973.

If the United States had nothing but oil shale to fall back on, there would be a decided energy shortage well into the 1990s. Even a free market could not prevent this shortage, in view of the long production cycle that, among other things, involves retraining hundreds of thousands of people, designing processing machinery that currently exists only in pilot-plant size, building living quarters and eventually communities replete with derivative manufacturing and service industries, not to mention environmental impact studies and court delays.

How much will it cost to produce shale oil? In 1971, the National Petroleum Council thought average production cost from Class 1 reserves would run $4-$5 per barrel. Today, $15-$20 per barrel seems a more reasonable estimate.

Like conventional oil operations, shale oil production entails high capital costs. In addition to mining, crushing, retorting and waste disposal, shale oil has to be upgraded in what might be called a pre-refining process that yields a product called syncrude, which is comparable to crude oil and can be processed in conventional refineries. Upgrading is required for at least two reasons: (1) high viscosity and the low pour point of shale oil (50°F) prevent shipping through pipelines to existing refineries; and (2) shale oil's high nitrogen content (1.7% by weight) would deactivate many of the catalysts used in refineries. The pour point of syncrude, by comparison, is 75°F; its nitrogen content is practically zero.

Unlike conventional oil operations, production of shale oil also entails very high variable costs. The mining operation necessarily requires an army of workers and huge quantities of materials, all contributing to variable costs. This makes shale oil production highly vulnerable to price reductions that might result from new domestic or foreign developments in the energy field. Whether U.S. oil companies bidding on oil shale leases have considered the inherent strategic implications of this characteristic is doubtful. In fact, environmental problems and highly uncertain economic prospects, brought about by congressional anti-developmental energy attitudes, forced the last U.S. oil company to withdraw from the shale-oil venture in November of 1976; pending an unlikely legislative reversal, oil shale ceased to be an energy alternative as of that date.

Tar Sands

Tar sands were once believed to be a ready alternative source of energy, especially since it was known that the Athabasca deposits in northern Alberta contained something like 800 billion barrels of bitumen capable of yielding 300 billion barrels of syncrude—80 billion by mining methods and 220 billion by thermal methods.[10] The United States, however, was pointedly reminded that this was foreign oil when Canada, acting under the OPEC price umbrella, raised the price of its crude oil.

That leaves strictly U.S. tar sands; in other words very little. Some 20 billion barrels of bitumen are known to be located in Utah, with additional small deposits scattered throughout the United States. The likelihood of large-scale U.S. tar sand exploitation by 1980 is remote, and syncrude produced from U.S. tar sands will never assume dimensions of national significance.

Synthetic Fuels From Coal

In discussing coal reserves, it is imperative to use a consistent probabilistic measuring rod. Thus, coal reserves are usually classified as measured, indicated, and inferred.

Measured reserves are thought to be accurate to within 20% of true tonnage; they are based on observation points, bore holes, etc., spaced not more than ½ mile apart. Indicated reserves are established on the basis of a 1 to 1½-mile grid pattern of bore holes. Inferred reserves, least certain of the three categories, are more interpretative and less definite.

Measured and indicated U.S. coal reserves total 394 billion short tons.[11] Of these, 261 billion tons (66%) are bituminous coal, 120 billion tons (30%) are subbituminous coal and lignite, with anthracite totaling only 13 billion tons, or less than 4%.

The reserve figure of 394 billion tons does not include thin beds, defined as beds with less than 28 inches of bituminous coal and anthracite, or less than 5 feet of subbituminous coal and lignite, nor does it include coal reserves at depths greater than 1,000 feet. These presently economically exploitable (measured and indicated) coal reserves of 394 billion tons represent some 25% of all thin bed, intermediate bed, and thick bed coal reserves (measured, indicated *and* inferred).

Assuming a 50% recovery efficiency of underground mining operations and 100% for surface mining, total recoverable coal reserves, based on the conservative 394-billion figure, amount to 220 billion tons. At the coal production rate in 1970, this corresponds to a supply life of well over 300 years. However, if coal is assumed to partially make up for the developing decline in crude production, a rapid growth in coal mining operations is sure to follow, and this will reduce the remaining life of coal reserves.

For example, a growth rate in coal production of 3%, compounded annually, corresponds to a reserve life of 85 years. If the growth rate rises to a phenomenal 5%, the measured and indicated reserves of 394 billion tons will last 60 years. Certainly, the United States is not going to run out of coal in this or the next generation, especially since the 5% growth rate considers use of coal for producing synthetic oil or gas. By comparison, the 1965 to 1970 growth rate of domestic coal consumption was 1.7% per year.

The future growth of coal consumption hinges on two new developments. First, an accelerated move towards direct use of coal in generating electricity, and second, synthetic fuels. Needless to say, each of these uses is beset with its own problems.

Coal-burning power plants are faced with stringent environmental control regulation, especially the Clean Air Amendments of 1970 (Public Law 91-604). The number one problem here is SO_2.

As an example of the magnitude of the problem, U.S. emission of SO_2 in 1970 was calculated to be 36.6 million tons, 20 million of which originated with coal- and oil-burning power plants. Respective projections for the year 2000 without abatement are 125.6 million tons total and 94.5 million tons for power plants. The only long-term answer to ever-tightening standards, as far as coal-burning power plants are concerned, is removal of SO_2 from combustion gases before dispersion. Removal of sulfur from fuel before burning is not feasible, since much of the sulfur is of the non-pyrite (organic) variety and not subject to mechanical separation.

Synthetic oil or gas made from coal are clean fuels, and some of them have useful applications beyond power generation, as automotive fuels. Essentially, two kinds of synthetic gases can be made using well-established processes:

1. A low-Btu gas for power generation;
2. Synthetic pipeline gas with a Btu content of 900-925 per scf, roughly 90% of natural gas.

Latest estimates place the production cost of high-Btu gas at $3 to $5 per Mcf, considerably above the regulated November 1976 price of $1.42 per Mcf of newly discovered gas. Thus, synthetic pipeline gas is a long way from becoming an economic reality.

No full-scale plant exists today in the United States for converting coal to syncrude. However, as in the SO_2-abatement area, many companies are doing research in this field and pilot plants are on the drawing boards. The cost of coal-base syncrude is currently estimated at $17 to $20 per barrel.

As in the shale-oil case, production of coal-base syncrude is subject to substantial research and development lags as well as to construction delays. One estimate projects total commercial syncrude production at 0.68 million bpd some 17 years after design of a prototype plant—that is 1% of the 1973 U.S. crude oil consumption. Again, the Heidelberg wine keg syndrome applies: huge reserves but a tiny spout. And again, the folly of the U.S. commitment to Project Independence commands the attention of any impartial and informed observer.

ii. **Non-Fossil Energy Alternatives***

Much has been written and said about non-fossil energy alternatives. Some of the more promising alternatives are the subject of the remainder of this chapter. It should be noted at the outset that these non-fossil alternatives generally are for the generation of electricity which, in 1974, represented no more than 27% of the total U.S. energy consumption. Thus, the non-fossil energy alternatives have severe limitations.

While it is conceivable, and in the long run inevitable, that the U.S. industrial and residential/commercial sectors will ultimately be converted to an all-electric mode of operation, such a conversion of the transportation sector is technologically not in sight. This latter sector represents 25% of the U.S. energy demand, and presently it is almost entirely oil-dependent.

The non-fossil energy alternatives that will be considered here are sources of solar, geothermal, and nuclear energy.

Solar Energy

Solar energy manifests itself in various energy forms such as radiant heat, winds, rainfall (which resupplies hydroelectric reservoirs), fossil fuels, organic farming, and ocean thermal gradients. In the past, solar energy indirectly supplied a major share of the energy used in pre-industrial and early industrial societies. As wind, it drove the mills, pumped water and sailed ships. After conversion to organic fuels (wood) by photosynthesis, it heated homes, schools, and factories and provided steam for vapor-cycle engines. As was pointed out in Chapter 1, the oil, gas, and coal which we use are really manifestations of accumulated solar energy stored in the earth over a period of about a billion years. At our projected rate of consumption, these fossil fuels, which required approximately one billion years to accumulate, will be used up in 200 years. No one knows for certain, but possibly 120 of those 200 years of fossil fuel supply have elapsed.

The different forms of fossil fuels will not be depleted at the same time. The most convenient forms will decline much faster than less convenient forms. One thing is certain, however: All fossil fuels are finite in volume and sooner or later our technology must be ready for non-fossil energy sources. The sun is one such source.

How much energy does the sun provide? Scientific measurements have shown that the amount of energy arriving from the sun at the earth's outer at-

*This discussion draws heavily from *Energy Alternatives: A Comparative Analysis,* U.S. Department of Commerce, NTIS, B-245 365, May 1975.

mospheric limit amounts to approximately 430 Btu/hr-ft^2. This unit of energy is known as the solar constant. Since the mean polar plane of the earth is approximately 49 million square miles, the total energy reaching the earth's outer atmosphere is 1.4 x 10^{19} Btu/day or 5.2 x 10^{21} Btu/year. Profiles of solar radiation, starting at the earth's surface and extending upward to outer space, indicate that approximately 30% of the outer atmospheric solar energy is immediately reflected back into space. Another 47% of our daily solar energy is absorbed into the atmosphere, leaving only 23% reaching the earth's surface. However, this 23% of the earth's total solar energy amounts to 3.3 x 10^{18} Btu/day or the equivalent of approximately 561 billion barrels of crude oil per day. This, in theory at least, is the amount of solar energy that is available for conversion through direct heating, ocean thermal gradient, photovoltaic cells, photosynthesis, and evaporation.

Serious proposals for converting solar energy to useful purposes have been along four general lines: the conversion of wind power, direct radiation (using space heaters at low temperatures, reflective solar boilers, or photovoltaic cells), ocean thermal gradients, and organic farms.

Wind Power

At present, there is no adequate basis for estimating the potential of wind power. This is due to the fact that the technology of convection currents has not been fully developed. Despite these problems, approximations are available for many areas of the United States.

About 2% of the solar energy absorbed into the earth's atmosphere is converted to wind energy. From this figure, it is calculated that the rate at which wind energy is being generated over the 48 contiguous states is about 14 times the 1973 energy demand.

Approximately 30% of the wind energy is generated in the lower 3,300 feet of the atmosphere. The wind energy available to man, however, is that which occurs in the first 500 feet from the surface. As the energy is removed from the wind close to the ground, kinetic energy is transferred downward from higher altitudes through convections in the earth's boundary layer. In a recent study sponsored jointly by the National Science Foundation (NSF) and the National Aeronautics and Space Administration (NASA), a research team estimated that an annual output of 5.1 x 10^{15} Btu of wind energy would be possible by the year 2000. That amount of energy represents 7.2% of the total U.S. energy demand in 1975, roughly the equivalent of the U.S. demand for electricity in 1972. The most promising geographical locations for wind power generation in the United States are along the coast lines and throughout the Great Plains.

The only method of conversion of wind power to useful energy is through windmills. For conventional windmills, the output from the rotor is a direct function of the square of the diameter of the blade and the cube of the wind velocity. It is this exponential relationship between wind velocity and output that places such a high premium on identifying sites with continuous high winds.

Small-scale applications of wind power are encouraging. A 10-foot rotor and an 8-knot wind 70% of the time can recharge the battery of a small urban car overnight. A 25-foot rotor under the same circumstances could supply all the electrical energy needed for a single family dwelling. Wind-power outages, however, cannot be avoided and must be planned for.

For large-scale applications, energy farms with generating units spaced over a grid offer some interesting possibilities. With this approach, power densities of 40 megawatts per square mile are possible in many places throughout the United States. Where the generating units are 200-300 feet high, conventional commerce, agriculture or urban dwelling can be supported by electrical energy derived from wind power.

The intermittence of wind energy is likely to be less critical than that of direct solar radiation. In fact, some combination of wind and solar radiation may be quite compatible. If both wind power and photovoltaic solar cells were developed over a multiregional power grid, the output from such a system could supply the base load, with conventional fossil-fueled units filling in to handle peak loads and power outages due to calm or cloudy days, as well as nighttime outages.

Direct Solar Radiation

Solar Concentrators. In low-temperature space heating, solar radiation can be captured on black surfaces which are covered with glass. When solar radiation falls on a darkened surface, the short-wave radiation is absorbed and converted into heat. The temperature of the surface will rise until it can dissipate energy at a rate in equilibrium with its absorption. Low temperature solar chambers have been demonstrated to reach temperatures between 225 and 250°F over a five-hour period during the high radiation portion of the day.

The design of the glass cover is an important feature of the radiation chamber. The incoming short-wave radiation must pass inward with ease but the long-wave radiation from the black surfaces must be restricted from passing outward. The glass cover also forms part of the convection chamber which moves the absorbed heat to the desired location.

To attain temperatures higher than 250°, the solar radiation must be concentrated through the use of reflecting surfaces. Highly polished parabolic silver surfaces, with piping carrying the heat-transfer medium running concentric with its focal lines, are relatively simple to construct. Temperatures on the order of 600°F are attainable with this type of unit. If such parabolic units are capable of tracking the sun, they can form the primary heat source for vapor-cycle engines. Several combinations of reflective solar boilers and vapor-cycle engines are in use today; however, they are not competitive with fossil-fuel power sources. The largest such installation today is in the Pyrenees Mountains in France. On a clear day, this installation can attain a thermal rating of 3.4 x 10^6 Btu's per hour and a temperature of 3,000°F.

The major advantage of extremely high-temperature solar concentrators is their potential for high conversion efficiencies in steam engines or steam turbines. Carefully designed heat absorbers located at the focal point are capable of heating flow-through working fluids to temperatures of 1,500°F or more.

Photovaltaic Cells. Photovoltaic power conversion involves the direct conversion of solar radiation to electricity through cleverly devised solid-state devices. A photovoltaic cell is constructed from semiconductor materials, such as silicon or germanium. Currently, silicon-based solar cells are available which convert sunlight to electric power with up to 18% efficiency. The cost of such cells is currently in the range of $20,000/kW. For practical application of photovoltaic cells, that cost would have to be reduced a hundredfold. Recent successes in the growth of continuous crystals and the abundance of the silicon base material offer some hope for achieving lower costs in the near future.

Several proposals have been made for the utilization of photovoltaic cells in the direct conversion of solar radiation to electrical energy. The most serious concepts are: solar farms in regions having high solar intensity, where relatively large areas are densely covered by photovoltaic cells; a solar strip several hundred feet wide running roughly along the 1,800 Btu/ft^2 per day contour shown on Figure 7-2; and a solar power satellite placed in an equatorial orbit around the earth.

The most promising geographic region for solar farms is the southwestern United States, where an area of approximately 6,500 square miles could produce all the electricity used in the United States in 1975. Still, many technical and legal problems have to be overcome before such a solar farm or even several smaller farms can be created.

The solar strip is actually a special configuration of a solar farm. However, since the solar strip would probably not be concentrated under a cloudy sky at any given time, its size can be reduced to as much as 4,000 square miles. Two such strips, half a mile wide and running from the Atlantic to the Pacific Coast, through the southern tier of the United States, could supply up to 75% of the total U.S. electricity requirement.

To avoid the losses due to atmospheric attenuation and nighttime outage, it has been proposed that large arrays of solar cells be placed in a satellite and sent into an equatorial synchronous orbit. The solar cells of this satellite would be exposed to the full intensity of sunlight 24 hours a day. The direct

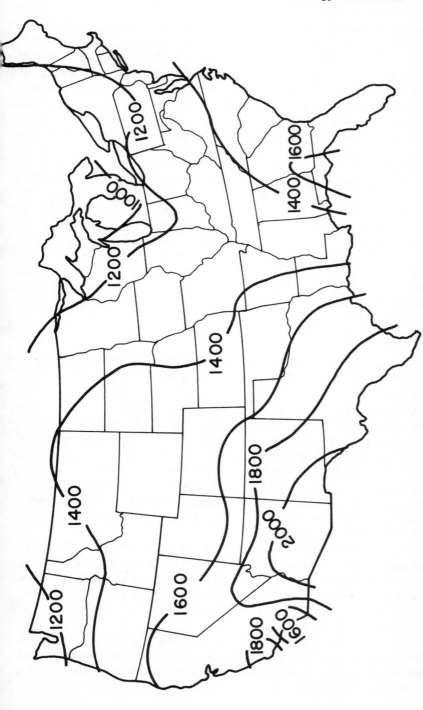

Figure 7-2. The distribution of U.S. solar energy. Figures give solar heat in Btu/ft.² per day. (Source: AEC, Draft Environmental Statement: Liquid Metal Fast Breeder Reactor Program, 1974.)

current produced by the photovoltaic arrays would be converted into microwave power and beamed to large receivers on the earth's surface. At the surface, the microwave energy would be reconverted back to electrical power. The concept envisions 25 square miles of solar cells in each satellite station and an area of 36 square miles for each ground receiver. It has been estimated that a satellite system of this size would provide 10,000 megawatts of electricity. The technical developments required to make satellite power stations feasible are so formidable that such stations are not likely to play any part in supplying energy in the foreseeable future.

Any viable solar energy system must be coupled with an appropriate energy storage system. The regular outage of solar energy requires a high degree of conversion efficiency both directly to usable power and to storable energy forms. During periods of outage, facilities must be available to convert the stored energy to on-line power. Storage systems which may be coupled to a solar source are: heat storage using molten salt, hydrogen, manufactured methane, impounded water, and batteries. The investment capital and operating expense required to provide solar energy on a 24-hour per day basis are immense to say the least. Much additional study will be required before such an energy system can become a realistic alternative.

Ocean Thermal Gradients

The temperature of the ocean near the surface in the regions between the Tropic of Cancer and the Tropic of Capricorn stays about 77°F. This uniform temperature is the result of cyclic solar heating balanced with heat losses from evaporation. At depths of 3,300 feet in these latitudes, the water temperature is 41°F. This temperature difference of 36°F could be the energy potential for generating electricity in a conventional vapor-cycle engine.

The economic development of ocean thermal power plants requires the use of a secondary fluid having a boiling point of 68°F. In operation, the working fluid would be heated and vaporized by the warm surface water in large heat exchangers. The vapor would then be expanded through turbines to produce electricity and finally condensed to the liquid state in heat exchangers located in the colder depths of the ocean.

The maximum theoretical thermal efficiency of an ocean thermal gradient power plant would be about 6.7%; the actual efficiency probably would not exceed 3%. Consequently, harnessing solar energy in this way is considered infeasible.

Organic Farms

During the pre-industrial era the supply of organic fuel (wood) was synonymous with the clearing of forests from lands which were desired for farming and ranching. In modern terms this activity would involve recycling of the land to preserve the energy-supplying potential of forests. This process is termed "organic fuel farming."

As mentioned earlier, the photosynthesis process is very inefficient. Coupled with the high costs of fertile lands, irrigation, fertilizer, land preparation, transportation, and labor, this places organic fuel farming beyond economic reach. As the population of the world increases, the demand for fertile land for food production will increase. With organic fuel farming impractical now, future developments will further compound its impracticality.

Geothermal Energy

Simply stated, geothermal energy is the earth's internal heat. While most of the earth's heat is contained in its core and mantle at depths unlikely to be penetrated by foreseeable drilling techniques, approximately 10^{24} Btu's of heat are stored at accessible depths. Some of the earth's heat has been concentrated in localized "hot spots" having temperatures ranging up to 650°F. These localized hot areas have been classified as dry steam, hot water, and hot dry rock systems. In these hot areas the heat flow from the core of the earth ranges up to five times the normal range.

For thousands of years man has used thermal springs to bathe and warm himself. Early records note the widespread use of natural hot waters in ancient Rome and in other European countries. In 1777, practical use was made of steam vents in Larderello, Italy. This city was also the site of the first power generating station using natural steam energy (1904). Since then, many countries have made practical use of naturally occurring geothermal energy.

The largest user of geothermal steam for heating is the Soviet Union, where savings in coal of 15 million tons per year have been claimed. Hungary and New Zealand also employ geothermal energy for heating. Geothermal power production, as of 1975, was ranked by country as follows: Italy number one, with 400 megawatts of installed capacity; the United States second, with 300 megawatts; New Zealand, with 200 megawatts; Mexico, 74 megawatts; Japan, 30 megawatts; and Russia, with 26 megawatts. Other countries with significant undeveloped geothermal potential are Algeria, Colombia, Chile, Dominica, El Salvador, Ethiopia, Guadalupe, Greece, India, Indonesia, Israel, Kenya, Morocco, New Hebrides, the Philippines, Rwanda, Taiwan, Tanzania, Turkey, and Uganda.

In the United States, interest in geothermal resources began in the early 1920s. Figure 7-3 shows the distribution of geothermal resources in the United States. In California, the geyser region is currently producing electricity for commercial use from geothermal steam. Two hundred homes

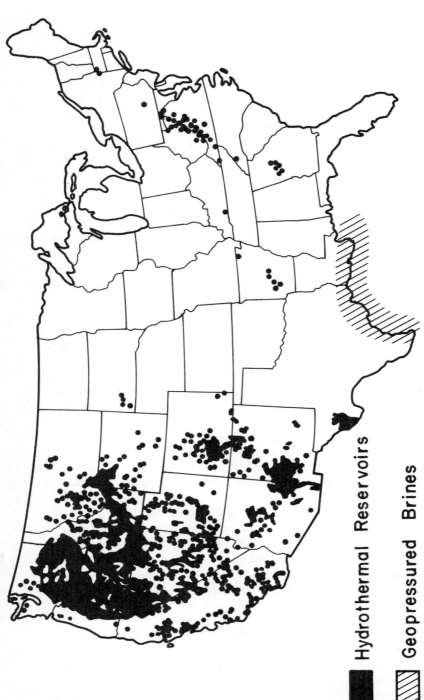

Hydrothermal Reservoirs

Geopressured Brines

Figure 7-3. The distribution of geothermal resources in the United States. (Source: U.S. Dept. of the Interior, Final Environmental Statement for the Geothermal Leasing Program, Vol. 1, pp. II-17, 1973.)

in Boise, Idaho, with usage dating from 1890, and 500 homes, 7 schools, and several factories in Klamath Falls, Oregon, are utilizing geothermal energy as a direct heat source.

Normally, the heat of the earth is diffuse. However, local geological conditions do exist, generally in areas of volcanic activity and in rapidly sinking geological basins, that concentrate heat energy into hot spots or thermal reservoirs. Three categories of thermal reservoirs are defined geologically as hydrothermal, geopressured, and dry hot rock reservoirs.

Hydrothermal reservoirs are the most adaptable geothermal reservoirs for producing energy. These reservoirs consist of a heat source (magma) overlain by a permeable formation (aquifer) in which the ground water circulates through the pore spaces. The aquifer is enclosed by impermeable boundaries which prevent water (steam) loss.

Water and steam transport the heat energy from the porous or fractured rock to a well which communicates to the surface. Two categories of hydrothermal reservoirs are recognized, based on the combination of water and heat content. Predominant vapor systems (high heat) such as the geysers in California are more attractive commercially, but such reservoirs are relatively rare. Reservoirs containing unusually hot water, sometimes classified as low-quality steam, are much more common.

Geopressured reservoirs differ from hydrothermal reservoirs mainly in their source of heat. Rather than deriving their energy from magma, the process involves clays in a rapidly sinking basin area, such as the Texas and Louisiana Gulf Coast. These clays effectively seal off the aquifer, thus trapping high-pressure, high-temperature water. The California Imperial Valley geothermal reservoir may be a combination hydrothermal-geopressured reservoir.

In dry hot rock reservoirs, no permeable aquifer is associated with the heat source. Consequently, this type of reservoir has no water or steam. The recovery of heat from this type of geothermal source requires the injection and circulation of extraneous water. Although several heat reservoirs of this type are known, to date none have been developed for commercial use.

The production system for geothermal energy is similar to oil or gas in that wells are drilled into the thermal reservoirs. These wells are cased and completed to provide a stable conduit for the hot fluid. All necessary equipment to produce and control the steam is located at the well head. Characteristically, steam cannot be transported over long distances (measured in a few hundred yards) because of the heat losses which would cause the steam to condense and lose pressure. Therefore, geothermal generating plants are necessarily close to the geothermal wells.

At the Geyser geothermal field in California, where the economics of geothermal utilization are well established, a typical generating plant has a capacity of 110 megawatts consisting of two 55-megawatt generators. To supply this plant, 15 geothermal wells were drilled, 14 of which are steam-producing wells and one is a condensate re-injection well. The wells are normally spaced about 50 acres apart so that a total of 800 acres, or more than one square mile, is required for each 110-megawatt generating station. The net generating capacity at the Geyser field in 1975 was 502 megawatts. Planned additions in the next five years will more than double its capacity, to 1,238 megawatts.

The Imperial Valley in Southern California is one of the largest and most thoroughly studied hot water fields in the United States. Currently, an experimental hot water facility with a potential capacity of 10 megawatts is testing the thermal effects of fluid re-injection into this reservoir. A recent estimate of the valley's thermal potential is 200 megawatts by 1985 and 1,000 megawatts by the year 2000.

The federal government, through the Energy Research and Development Administration (ERDA), has estimated recoverable energy from geothermal resources in the United States along with their status of technological development. The results of ERDA's study are shown in Table 7-3.

Table 7-3
Geothermal Energy

Resource Type	Billions of Barrels of Oil Equivalent		State of Technology
	Known	Inferred	
Dry steam	0.4	0.4	Commercial
Hot water	18.3	65.7	Test phase
Geopressured zones			
Electric utilization	18.3	42.0	Experimental
Methane production	91.3	273.8	Experimental
Hot dry rocks	14.6	43.8	Experimental
Total	142.9	425.7	

Estimates of future geothermal developments in the United States in terms of electrical capacity vary widely. Industrial planners forecast that the geothermal system will provide 3,000 megawatts of power by 1985. This is less than 1% of the U.S. generating capacity in 1925. By 1990, the electrical output from geothermal energy is expected to double to 6,000 megawatts, still according to the views of industrial planners. The federal government planners have a more optimistic view.

Nuclear Energy

The first nuclear power plant went into operation in 1957 in Shippingport, Pennsylvania. Just 12 years after the two nuclear blasts in Japan, through the exploits of the Atomic Energy Commission (AEC), a weapon of war was modified for peaceful use. The complexity of nuclear technology plus the associated hazards of radiation and criticality (nuclear detonation) made its cost greater than private industry was willing to bear at the time. Consequently, the transition of nuclear technology from military to commercial use was accomplished by the United States government.

Nuclear fission is the process whereby certain radioactive atoms are split into two dissimilar atoms and one or more neutrons. In the process of splitting the primary atom a great amount of energy is released. The neutrons which are released collide with other primary atoms, causing them to fission, thus creating a chain reaction. The technology which controlled nuclear criticality at manageable rates was the breakthrough which led to the commercial application of nuclear energy.

Radioactive isotopes of uranium and plutonium which fission readily are referred to as fissiles. Three such isotopes are U-233, U-235 and Pu-239. When an atom fissions, the two newly formed atoms are called fission products or fragments. Products of fission are strontium, cesium, iodine, krypton and xenon. These products are transformed to stable products over time through radioactive decay. In the process of decay, there is a concurrent release of energy in the form of highly energetic alpha, beta and gamma particles. These particles can have beneficial uses (including several in medicine), but they can also have adverse effects on the cells of biological organisms.

In the 30 years of applied nuclear fission research, two generations of technology are either available or under development. They are conventional fission reactors and breeder reactors. In May of 1974, conventional fission reactors generated 27,800 megawatts, or 6.1% of the electrical power used in the United States at that time. Two types of conventional fission reactors are presently available. They are the light-water reactor (LWR) and the high-temperature gas reactor (HTGR). Over 50 such reactors are currently operational in the United States. By 1980, the AEC estimates that nuclear reactors will furnish approximately 16% of the total U.S. electrical power.

The concept of breeder reactors developed after it was discovered that one of the products of fission is Pu-239, which results from bombarding U-238 with neutrons. The liquid metal fast breeder reactor (LMFBR) is currently under development, but the AEC does not forecast its commercial availability in the United States until around the year 2000.

Coal is one of the major fuels used in the generation of electricity. For perspective, one pound of uranium can produce as much energy as 3 million pounds of coal, assuming coal has a heating value of 10,000 Btu/lb. This same amount of energy is found in 5,155 barrels of crude oil.

The following discussion of nuclear reactors assumes a 1,000-megawatt nuclear plant operating for a period of one year at a load factor of 80%.

Light-Water Reactor (LWR) System

The light-water nuclear reactor gets its name from the fact that it uses ordinary water as the heat-transfer medium from the fission area to the steam turbines. The primary fuel source for the LWR is U-235. The fuel cycle of the LWR contains ten major steps, starting with the exploration for uranium and progressing through mining, the milling of U_3O_8 or yellow cake, the production of uranium hexafluoride (UF_6), enrichment of U-235, fuel fabrication, the fission to produce electricity, reprocessing of the used fuel to recover the unspent U-235 and the newly created Pu-239, waste management, and transportation and storage of nuclear fuels and by-products. A diagram of the fuel cycle is shown in Figure 7-4.

Uranium consists of three naturally occurring isotopes in the following proportions: 99.29% U-238, 0.71% U-235, and a trace of U-234. The fuel for the LWR is U-235. A ton of uranium ore generally contains 4 to 5 pounds of uranium oxides but only 0.35 to 0.48 ounces of U-235.

Uranium reserves are normally discussed in terms of quantities available at three cost-of-recovery levels of $8, $10, and $15/lb of U_3O_8. In 1974 the AEC made the following estimates of reserves of uranium at these three prices:

U.S. Uranium Reserves

Production Cost (1974 $/lb)	Resources (thousands of tons of U_3O_8)	
	Reserves	**Potential**
8	227	450
10	340	700
15	520	1000

In recent months the market value of yellow cake (U_3O_8) has zoomed past $50/lb because it is an alternative to the rising cost of crude oil. As a note of interest, the free world reserves of U_3O_8 at $15.00/lb were estimated to be 1.2 million tons.

A typical 1,000-megawatt LWR requires 200 tons of yellow cake (U_3O_8) per year. Therefore, the presently licensed capacity of approximately 28,000 megawatts would exhaust the nation's $8/lb reserves in about 49 years. If the United States achieved the 250,000-megawatt capacity projected by the AEC for 1985, existing $15/lb reserves would last only ten years. Table 7-4 summarizes the projected needs for U_3O_8 to supply LWR's.

At present (1977) there are two different types of LWR's: the boiling water reactor (BWR) manufactured by General Electric, and the pressurized water reactor (PWR) manufactured by Babcock and Wilcox, Combustion Engineering, and Westinghouse. In all LWR's the heat comes basically from the fissioning of U-235 atoms. The fuel cells in LWR's generally contain 97%

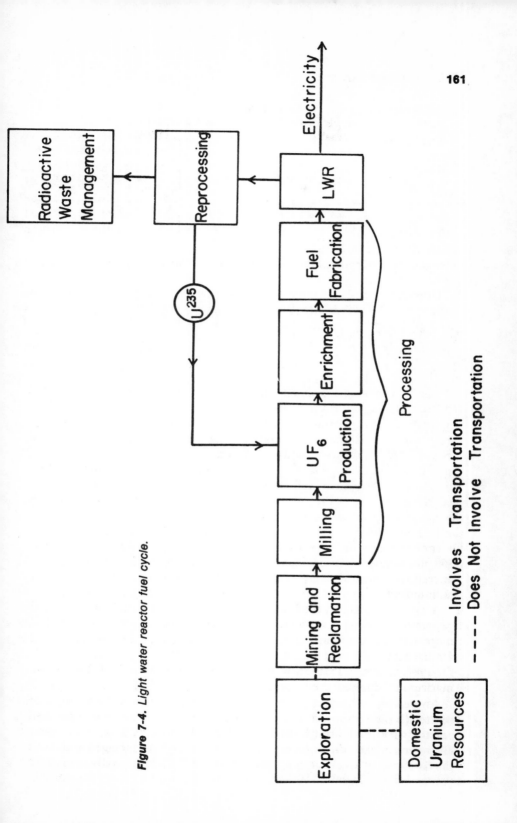

Figure 7-4. Light water reactor fuel cycle.

Table 7-4
U₃O₈ Needs for Projected Light-Water Reactor Capacity

Date	AEC Projected Nuclear Capacity (Mwe)[a]	Tons of U_3O_8 Needed per Year	Number of Years the Proven Reserves Will Last at the Given Nuclear Capacity		
			$8/lb.	$10/lb.	$15/lb.
1974	28,183	5,367	49	60	92
1980	102,000	20,400	13.5	16.5	25.5
1985	250,000	50,000	5.5	6.8	10.4

[a] Source: INFO, Public Affairs and Information Program, No. 70, Atomic Industrial Forum, Inc., May 1974.

U-238 and only 3% U-235. In the fission reaction some of the particles released convert the U-238 to Pu-239. For each gram of U-235 consumed, approximately 0.6 gram of Pu-239 is formed. Generally, more than half of the new plutonium undergoes fission in the reactor core, thus contributing significantly to the energy produced in the power plant.

Figure 7-5 shows a simplified schematic of a boiling water reactor. In this version, water is pumped in a closed cycle from the condenser to the nuclear reactor core. Heat generated by the fissioning uranium pellets in the reactor core is transferred through the metal cladding to the water flowing around the fuel assembly. As the water boils, a mixture of steam and water flows out the top of the core through a steam separator to the turbine-generators. The turbine exhaust is condensed and returned to the reactor pressure vessel to complete the cycle. The BWR system is termed a "direct-cycle" system because the energy supplied to the water from the hot fuel is transported directly to the turbine. The pressure in a typical BWR is maintained at about 1,000 pounds per square inch (psi) with a steam temperature of 545°F. Neutron-absorbing control rods, operated by hydraulic drives located below the core, are used to control the rate of the fission chain reaction.

It is important to realize that a nuclear reactor cannot explode like a bomb. The type of fuel and fuel configuration used in a reactor is different from that of a bomb. The most hazardous event that can happen in an LWR is the disruption of the circulating water, which would cause the core to overheat and melt down.

The pressurized water reactor (PWR) is shown schematically in Figure 7-6. The primary difference between a PWR and a BWR is that all PWR's employ a dual coolant system for transferring energy from the reactor systems. In the PWR, the primary loop is filled with water that is pumped around the core and through the heat exchanger. The secondary loop is water that is pumped past the primary heat exchanger to the turbine-generators. In the core, the primary water is heated to about 600°F, but the pressure is kept sufficiently high (about 2,250 psi) to prevent vaporization. After the water in

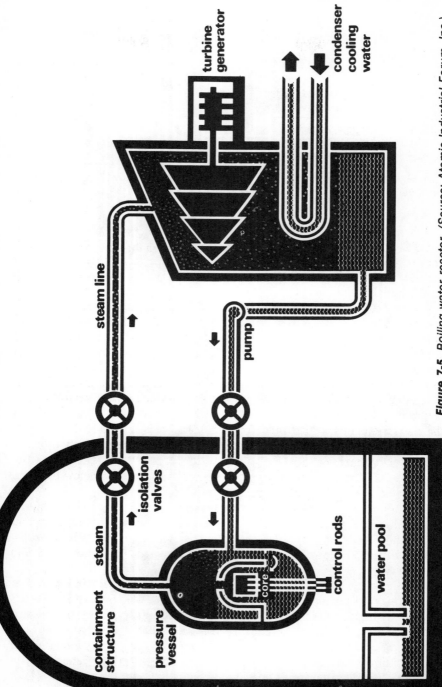

Figure 7-5. Boiling water reactor. (Source: Atomic Industrial Forum, Inc.)

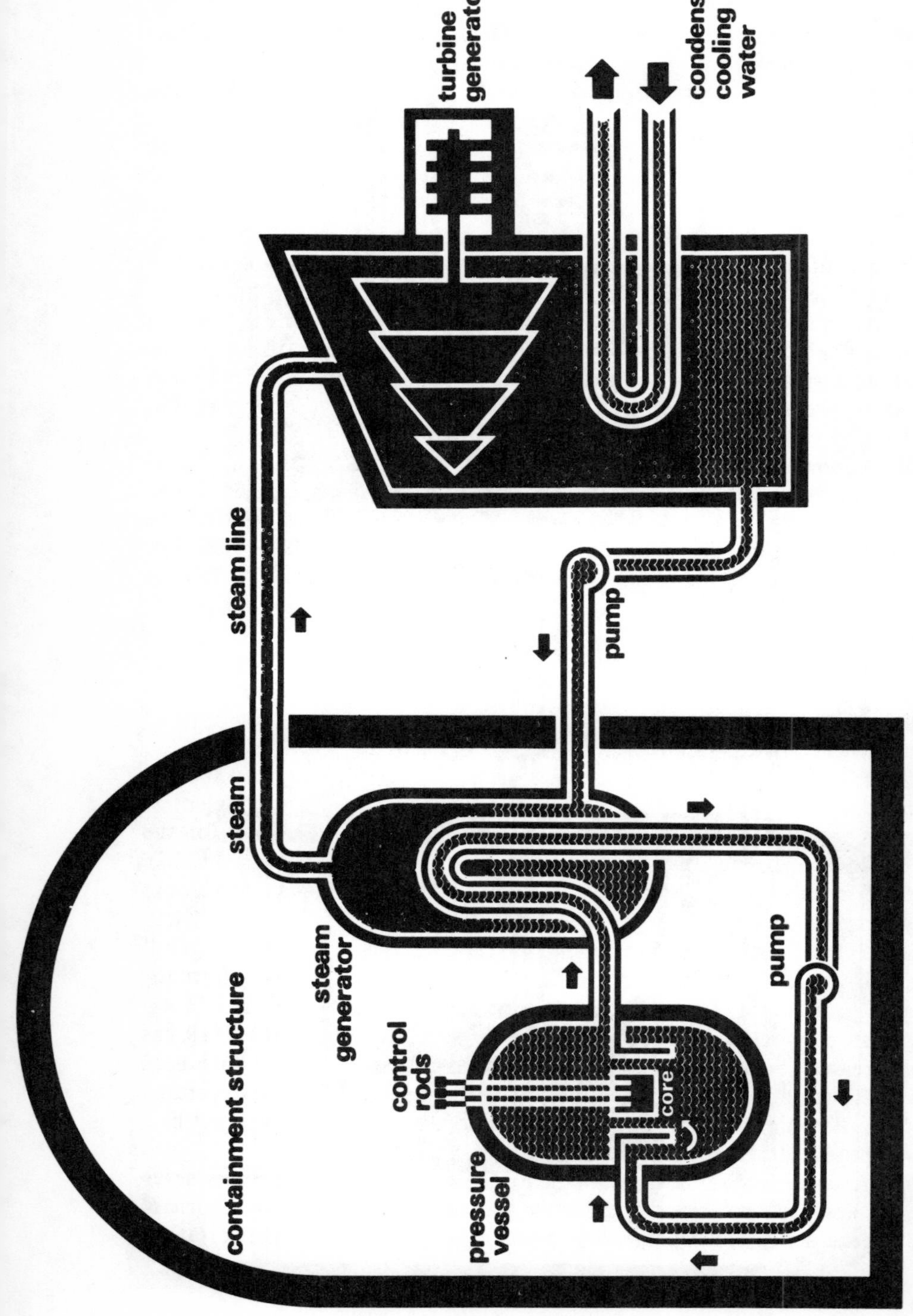

Figure 7-6. Pressurized water reactor. (Source: Atomic Industrial Forum, Inc.)

the secondary loop passes through the condenser, downstream from the turbine, it is returned to the primary heat exchanger.

The overall energy efficiency for the LWR's (both BWR and PWR) is the ratio of electric energy output to total fission energy released. With the LWR's, an efficiency of 32% is achieved compared to 38 to 40% for modern fossil-fueled power plants. The reason for this lower efficiency is that LWR plants can only operate at a maximum steam temperature of about 600°F, while fossil plants can operate at 1,000°F or higher.

The cost of providing the fuel system for a LWR comprises the total cost of the fuel cycle shown schematically in Figure 7-4.

Assuming an 0.2% U_3O_8 concentration in the ore, 100,000 tons of uranium ore would have to be mined each year (200 tons of yellow cake) to supply one 1,000-megawatt LWR. For comparison, a 1,000-megawatt coal-fired plant would require approximately 3 million tons of coal per year, assuming that the heating value of coal is 10,000 Btu's per pound.

The economics of the BWR and PWR are very similar. Their capital costs (in 1974 dollars) have been projected to be in the range of $411 and $472/kW installed capacity for a plant to begin operation in 1981. The capital cost for a coal-fired plant of similar size would range between $386 and $444/kW. For an oil-fueled plant, the capital cost would range between $280 to $334/kW.

High Temperature Gas Reactor (HTGR) System

Helium is used as the coolant and heat-transfer medium in this type of reactor; hence the name "high-temperature gas reactor" (HTGR).

The HTGR differs from the LWR's both in heat efficiency and fuel requirements. The capacity to heat helium to a higher temperature at high pressures allows the HTGR to achieve efficiencies of 40%. The fuel for the HTGR is composed of thorium (Th-232) and uranium (U-233 and U-235). The fuel in the HTGR is formed into microspheres and is molded into blocks by a graphite binder.

The HTGR was developed mainly as a commercial venture by Gulf General Atomic Corporation and received limited financial support from the federal government. The only commercial HTGR in operation is in a 40-megawatt plant at Peach Bottom, Pennsylvania. A 330-megawatt HTGR has been licensed for operation in Fort St. Urain, Colorado. Orders have been placed for ten additional units, but future roles of the HTGR are uncertain. In 1973, Battelle Research Institute estimated that the HTGR will not be a major producer of electrical power until the year 2000.

The resource system diagram for the HTGR (Figure 7-7) includes seven major steps: exploration, mining for and processing of uranium and thorium, power generation, fuel reprocessing, waste management, and transportation.

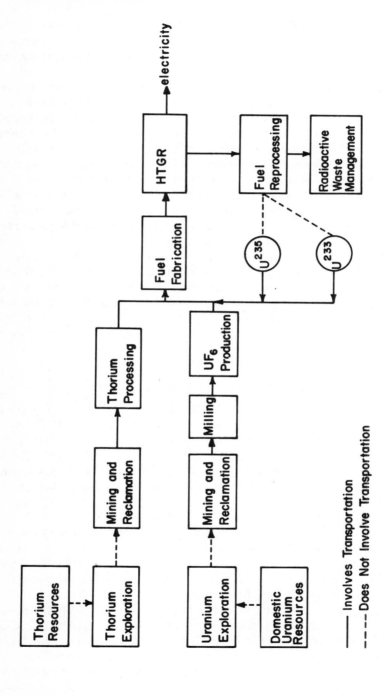

Figure 7-7. High-temperature gas reactor fuel cycle.

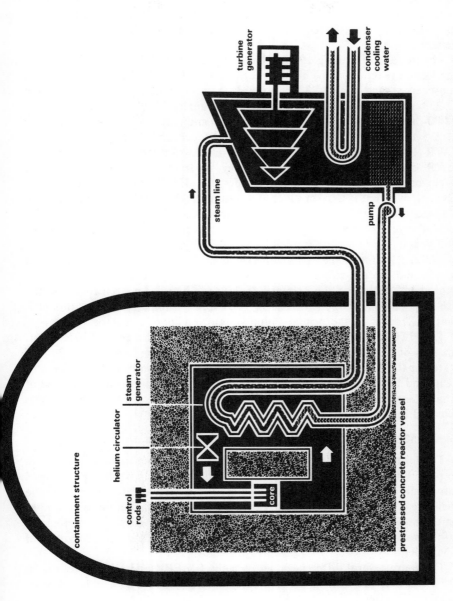

Figure 7-8. High-temperature gas-cooled reactor. (Source: Atomic Industrial Forum, Inc.)

Figure 7-8 is a schematic of the HTGR. The heat created by the fissioning of U-235 and U-233 is transferred to the helium which circulates around the core. The helium is then circulated through a heat exchanger where the heat in the helium is transferred to water, making steam, which drives the turbine-generators.

The initial fuel loading of the HTGR core consists of U-235 and Th-232. In the fission reaction, the Th-232 is converted to U-233. The used fuel will be reprocessed and the U-233 recovered. After start-up, the U-235 and the U-233 will become the regular fuel. The annual fuel requirements for a 1,000-megawatt HTGR are 13 tons of U_3O_8 and 8 tons of thorium oxide. The U-235 used in the HTGR differs in concentration from that used in the LWR. U-235 for the HTGR must be enriched to 95%, versus 3 to 4% for the LWR. Th-232, a basic element, is a heavy, silvery metal which has numerous occurrences in the world. Thorium for the HTGR comes mainly from Canada, where it is a by-product of uranium processing. Thorium reserves in Canada are estimated as sufficient to supply the HTGR for 100 years. The U.S. thorium reserves, at a price of $50/lb (1973), if developed, would last 400 years.

Major features of the HTGR are the control rods that control the rate of fission and the pre-stressed concrete reactor vessel (PCRV). All the major equipment, including the primary coolant (helium), is contained inside the pre-stressed concrete reactor vessel. The PCRV eliminates the risk of a leak that may occur in the primary pipe system outside the reactor vessel.

The HTGR has a number of other safety advantages over the LWR: First, the loss of helium coolant is not a serious problem compared to the loss of water in the LWR; second, the graphite core can absorb large amounts of heat, and the temperature change of the core will be slow; third, the use of small, coated fuel particles reduces the amount of radioactive fission products released into the coolant.

Liquid Metal Fast Breeder Reactor (LMFBR) System

The term "fast breeder reactor" refers to nuclear reactors that, in addition to providing heat for the generation of electricity, also convert surplus U-238 into fissile Pu-239. Through the conversion of U-238 to Pu-239, more fissile fuels are created than consumed. The AEC had been conducting basic studies on the breeder reactor for some 20 years. About eight years ago an intensive effort was launched, in cooperation with industry, to develop a commercial LMFBR.

There are several differences between the LMFBR and other reactors. In the LMFBR, the central reactor core is surrounded by an outer core or blanket. The fuel rods in the central core contain a mixture of plutonium oxide (PuO_2) and UO_2, primarily U-238, while the blanket is filled only with UO_2. As the plutonium in the central core fissions, neutrons interact with the

U-238 in both the core and the blanket, transforming the U-238 into Pu-239. For every 4 pounds of Pu-239 consumed in the LMFBR reactor, approximately 5 pounds of new Pu-239 will be created in the core and blanket, thus the term "breeder" reactor.

The LMFBR has other distinctive features compared to LWR's and HTGR's. First, the LMFBR employs "fast" neutrons to achieve fission. Other reactors employ a moderating substance, such as hydrogen, to slow the speed of the released neutrons. Second, the LMFBR uses liquid sodium to transfer heat from the core to the water heat exchanger. Basic advantages of the LMFBR are: (1) the liquid metal coolant permits higher operating temperatures, thus giving projected plant efficiencies of 41%; and (2) the creation of fissile fuel (Pu-239) out of U-238, thereby greatly increasing usable nuclear energy resources. There are also disadvantages: Plutonium is one of the most toxic materials known, and sodium is extremely reactive chemically, requiring special significant safety precautions.

A simplified flow diagram for the LMFBR fuel system is shown in Figure 7-9. U-238 and Pu-239 are both involved in the fuel fabrication step. The three U-238 supply options are: uranium, which is mined and processed as in the LWR without the enrichment step; U-238 from the depleted stream in the uranium enrichment step for the LWR systems; and U-238 that is recovered from the used LWR fuel. Plutonium comes from two sources: Pu-239 recovered from the used LWR fuel, and Pu-239 that is bred in an LMFBR.

The principal effort in the LMFBR development in the United States is the Fast Flux Test Facility (FFTF) located at Oak Ridge, Tennessee. While the LWR system uses the U-235 isotope, which constitutes only 0.71% of naturally occurring uranium, the LMFBR utilizes the remaining U-238 isotope, which constitutes 99.29% of the resource. Thus the total energy resource base for the LMFBR is many times larger than the LWR energy resource base.

The LMFBR system will require an initial charge in order to operate until the self-generated plutonium supply is sufficient. This initial plutonium must come from the LWR system. Plutonium supplies will be made available through the LWR system, by the growth rate and timing of the LMFBR economy, and by the doubling time of the plutonium inventory due to the breeding gain in the LMFBR's themselves. Figure 7-10 shows a projection of plutonium availability and requirements. Based on these projections, the Pu-239 requirement for LMFBR's will not exceed availability from the LWR's until the year 2000. By that time the LMFBR's will provide inventory for new plants.

Figure 7-11 is a schematic diagram of the LMFBR power generation system. The central core contains mixed oxide fuel rods, while the blanket fuel rods are loaded with UO_2 pellets only. Fast neutrons created during operations convert the U-238 in both the core and the blanket to plutonium.

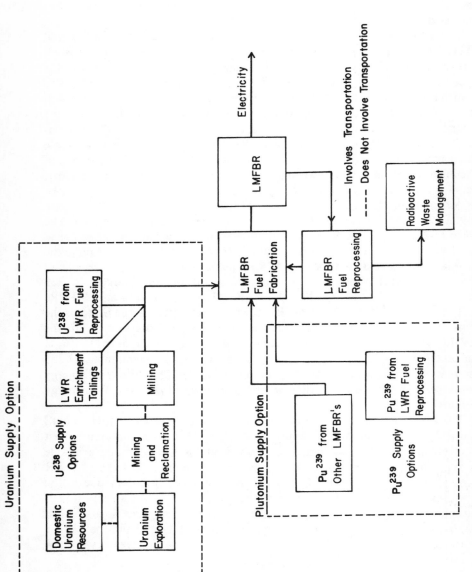

Figure 7-9. Liquid metal fast breeder reactor fuel cycle.

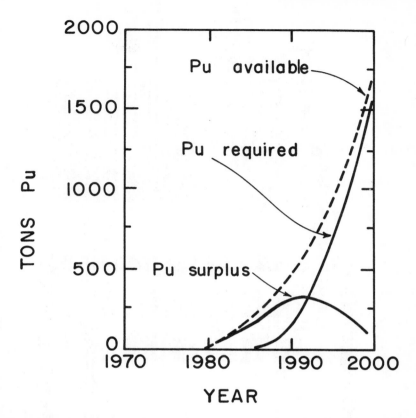

Figure 7-10. *Plutonium availabilities and requirements. (Source: R.J. Creagan, "Boon to Society: The LMFBR," Mechanical Eng'g., Feb. 1976.)*

Heat created by the fissioning fuel is absorbed by liquid sodium surrounding the core. The sodium in the primary loop flows through a heat exchanger where its heat is transferred to sodium in a secondary loop. This is necessary due to the tendency for sodium to become radioactive around the reactor core. The heat in the secondary sodium loop is then transferred to a water- or steam-cycle loop which drives the turbine-generators. The sodium used in the primary and secondary loops is an alkali metal that melts at 210°F, vaporizes at 1,640°F and has excellent heat-transfer properties.

At this stage of development, LMFBR cost estimates are speculative at best. The AEC has estimated the cost of LMFBR power plants in comparison to the LWR's and HTGR's. Table 7-5 shows these comparisons (1974 dollars were used for all estimates).

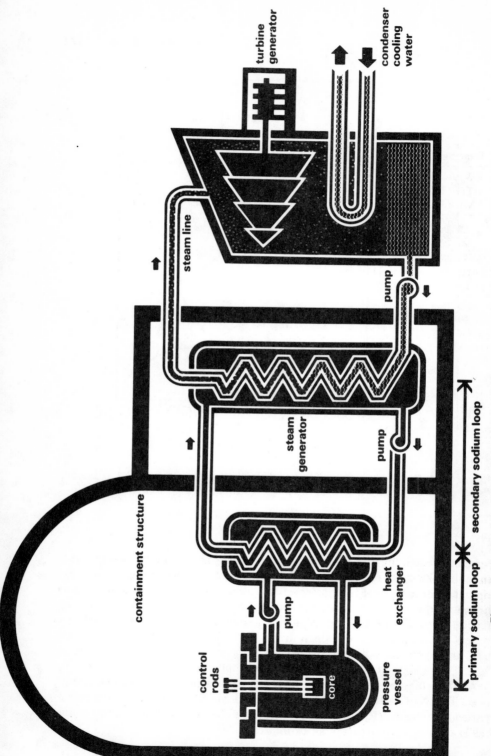

Figure 7-11. Liquid metal fast breeder reactor. (Source: Atomic Industrial Forum, Inc.)

turbine generator

condenser cooling water

steam line

pump

pump

steam generator

heat exchanger

pump

control rods

core

pressure vessel

containment structure

primary sodium loop

secondary sodium loop

Table 7-5
Estimated Nuclear Power Plant Capital Costs
(1974 dollars)

Reactor	Capital Cost ($/kWh)		
	1974	1980	1990
LWR (1,300 MW)	420	420	420
HTGR (1,300 MW)	419	419	419
LMFBR (1,300 MW)	NA	NA	487

Breeder reactors are not unique to the United States. Many other in-dustrial nations have placed the LMFBR on a high priority and are presently developing their programs more rapidly than the United States. For example, the U.S.S.R. is well along in the construction of a 150-megawatt demonstra-tion plant. Britain began operation of a 250-megawatt plant in 1972. France started up a similar plant in 1973, and Japan has announced plans to con-struct and start up a commercial-size plant before 1980. A utility made up of French, Italian, and West German interests has announced plans to purchase two 1,000-megawatt LMFBR's, one to be located in France and the other in West Germany. Also, West Germany, Belgium, and the Netherlands have organized to fund construction of an even larger unit with 2,000-megawatt capacity.

Heavy-Water Reactors

Water in which the hydrogen atoms contain a neutron in addition to the normal proton is termed "heavy" water. Such hydrogen molecules are called deuterium. Heavy water has desirable effects on the nuclear reaction; therefore, it is used as the heat-exchange medium in the primary loop of a nuclear reactor. As in the pressurized water reactor (PWR), heat is trans-ferred to a secondary system which is filled with normal water. Steam created in the secondary loop drives the turbine-generators. Heavy-water reactors are operated on natural uranium, whereas light-water reactors require enriched uranium. Heavy-water systems were developed and designed primarily in Canada, and the United Kingdom has opted for these reactor systems.

The Nuclear Energy-Fusion Resource System

The first demonstration of fusion as a source of energy was the explosion of a hydrogen bomb in 1972. Since that time, the AEC has had a continuous research project with the goal of developing technology for controlling fusion for the generation of electrical power.

Fusion reactions require the use of heavy isotopes of hydrogen, such as deuterium and tritium. Deuterium exists naturally in seawater, while tritium can be manufactured from lithium in the normal operations of a fusion reactor. Lithium is a relatively abundant element which is easily produced.

Environmentally, fusion reactors are expected to be more attractive than fission because of less serious fuel handling problems, lower radioactivity inventories, and because fewer radioactive wastes will be generated.

The development of fusion reactors has moved along two concepts. The first is magnetic confinement, in which the hydrogen isotopes exist in a plasma (gaseous phase) that is contained within a magnetic field. The isotopes are accelerated to ultra-high velocity and caused to collide. Fusion results from the collision. The second is the laser concept, in which concentrated light from lasers is used to compress and heat a pellet of hydrogen isotope. Collisions resulting from this activation cause fusion.

The development program being carried out by the AEC contains five major steps in magnetic containment technology, ending with the construction of a demonstration plant. Present estimates suggest that commercial use of fusion energy is at least 25 years away.

References

1. American Petroleum Institute, American Gas Association and Canadian Petroleum Association, *Reserves of Crude Oil, Natural Gas Liquids and Natural Gas in the U.S. and Canada,* 1970, p. 18.
2. Gardner, F.J., "1973: The Year of Major Changes in Worldwide Oil," *The Oil and Gas Journal,* Dec. 31, 1973, pp. 83-88.
3. *The Petroleum Situation,* The Chase Manhattan Bank, Feb. 28, 1974, p. 1.
4. Oil-in-place, excluding North-slope, from *Petroleum Facts and Figures,* 1971 Edition, American Petroleum Institute, p. 115, and from *API Annual Reserve Estimates, 1972.*
5. For some technical descriptions of various miscible floods, see for example: Fitch, R.A., and Griffith, J.D., "Experimental and Calculated Performance of Miscible Floods in Stratified Reservoirs," AIME Transactions, 1964, No. 231, pp. 1289-1298. Kloepfer, C.B., and Griffith, J.D., "Solvent Placement Improvement by Pre-Injection of Water, Lobstick Cardium Unit, Pembina Field." SPE Paper 948 presented at SPE 39th Annual Fall Meeting, Houston, Oct. 11-14, 1964. Blackwell, R.J., and others, "Recovery of Oil by Displacement with Water-Solvent Mixtures," AIME Transactions, 1960, No. 219, pp. 293-300. Caudle, B.H., and Dyes, A.B., "Improving Miscible Displacement by Gas-Water Injection," AIME Transactions, 1958, No. 213, pp. 281-284. Yarborough, L., and Smith, L.R., "Solvent and Driving Gas Compositions for Miscible Displacement," *Society of Petroleum Engineers Journal,* Sept. 1970, pp. 298-310. Ballard, J.R., and Smith, L.R., Reservoir Engineering Design of a Low-Pressure Rich Gas Miscible Slug Flood," *Journal of Petroleum Technology,* May 1972, pp. 599-605.
6. Hardy, W. C., "Deep-Reservoir Fireflooding Economics for Independents," *Oil and Gas Journal,* Jan. 18, 1971, p. 60.

7. *Heavy Crude Oil*, U.S. Bureau of Mines Information Circular 8352, p. 7.
8. *Wall Street Journal*, Jan., 1974, p. 1.
9. For a very good discussion of the ownership question, see Ellis, T.J., "The Potential Role of Oil Shale in the US. Energy Mix: Questions of Development and Policy Formulation in an Environmental Age, Ph.D. dissertation, Colorado State University, 1972, University Microfilms Catalog No. 73-13,030.
10. Humphreys, R.D., "Pioneer in Oil Sands Development," 1973 Symposium on Oil Sands, Fuel of the Future, held in Calgary, Alberta, Sept. 5-9, 1973.
11. These and most of the other statistical data shown in this section are from *U.S. Energy Outlook, Coal Availability*, National Petroleum Council, 1973.

8
Energy
Policies

i. President Ford's Energy Policy

U.S. energy policy is made on at least two levels: in the executive branch, ultimately represented by the President of the United States, and in Congress. Each policy-making unit has its own preferences, yet neither of the two units can function without the cooperation of the other. For example, the White House can devise a plan to intensify oil exploration efforts through tax incentives, but the Congress has to approve these through appropriate legislation. Similarly, congressional legislative measures need the President's signature to become law.

Presidential economic plans have never been accepted by the Congress as proposed, particularly not in the recent past when a Republican President was confronted with a Democratic majority in the Congress. Thus, presidential policy proposals only point in the direction which one participant in the policy-making process would like to take, and that is rarely the ultimate course that the nation's economy will follow.

In January of 1975, President Ford had his first real opportunity to cover energy policies in his State of the Union Address. The oil embargo was over, even though world crude oil prices remained at a level roughly three times as high as pre-embargo domestic oil prices. The Federal Energy Office and its successor, the Federal Energy Administration, had been created and had had time to study the U.S. energy situation. Indeed, their *Project Independence Report* of November 1974 was in the President's hands to help him shape U.S. energy policies. What, then, were the President's policies as announced in the 1975 State of the Union Address, and where did they lead the U.S. economy?

Table 8-1
The Energy Conservation Tax

	Year	Daily Production	Annual Production	Unit Tax	Total Tax (billions)
Gas	1974	61.6×10^6 Mcf	22.5×19^9 Mcf	$0.37/Mcf	$ 8.0
Domestic Oil	1974	11.0×10^6 bbl.	4.0×10^9 bbl.	$2.00/bbl.	$ 8.0
Imported Oil	1974	6.0×10^6 bbl.	2.2×10^9 bbl.	$2.00/bbl.	$ 4.4
				Total	$20.4

The most encouraging part of the President's program was the proposed deregulation of *new* natural gas (it would have been better to extend that to all gas) and the proposal to decontrol the price of domestic crude oil, presumably all of it, on "Presidential initiative." Both measures would have provided new incentives to oil companies to drill the sorely needed and highly expensive wells on the outer continental shelf and in Alaska, but neither of these measures survived the legislative follow-up to the President's State of the Union Address.

The President also proposed to couple this price increase with additional long-term taxation in the amount of $2 per barrel of domestic and imported oil and 37¢ per Mcf of gas. This so-called energy conservation tax was said to produce $30 billion per year, but the calculations in Table 8-1 shed doubt on that.

Making allowance for the then projected reduction in oil imports by one million barrels per day by the end of 1975 (which never did materialize) the total tax take shown in Table 8-1 would have been reduced by $0.4 billion and thus would have come out at exactly $20 billion. This means that the so-called windfall profits tax must have been relied upon to produce another $10 billion.

Since any tax increase is ultimately funded by the consumer, it is more than an academic question to ask how it would have affected the price of gasoline. The answer is that gasoline prices would have risen by 7.5¢ per gallon:

Revenue from energy conservation tax
Oil ....$12 billion
Gas ....$ 8 billion

Total.......................................$20 billion

Increase in gasoline prices due to above excise tax on oil, assuming that tax is spread evenly on all crude-oil
products.. 5¢/gal.

Required windfall-profits tax to fund the $30 billion
program...$10 billion

Increase in gasoline prices due to windfall-profits tax, assuming oil and gas shares that tax in the same proportion as the excise tax .... 2.5¢/gal.

Total tax-induced increase in gasoline prices .... 7.5¢/gal.

Incredibly, the annual tax receipts in the amount of $30 billion were not specifically earmarked for expanded U.S. energy research and exploration. Somewhat vaguely, they were to be returned to the economy "to compensate for higher fuel costs." But why would anyone tax an industry, thereby driving up prices, and then redistribute these taxes to compensate for higher prices?

This part of President Ford's energy message made little sense: Faced with a national energy shortage, the energy sector was hit by a $30 billion tax, and most of the funds generated through that tax were to be diverted to the economy in general. And what a tax! When the energy sector needed all the funds it could get its hands on, what it really got was a proposed tax bill three-quarters the size of the total net corporate taxes paid in fiscal 1974 by all the U.S. corporations taken together.

If the tax revenues that were to be generated from the energy sector had been slated for reinvestment in that sector, the policy proposal would have made sense, since this would have been a positive, even if not the most efficient, mechanism to develop new sources of energy. Short of such a reinvestment of energy taxes, another $20-30 billion would have been needed, through private or public channels, to implement President Ford's 1975 energy plan which included the construction of 200 major new nuclear power plants, 250 major new coal mines, 150 major coal-fired power plants, 30 major new refineries, 20 major new synthetic fuel plants, not to mention the drilling of thousands of new oil wells and the insulation of 18 million homes—all this over the next 10 years.

The additional investment needed for the implementation of these ambitious plans corresponds to 5-7.5¢/gal. of gasoline, so that the overall cost of President Ford's 1975 proposal would have resulted in an increase in gasoline prices by 12.5 to 15.0¢/gal., for a post-embargo price of 65.5 to 68.0¢/gal. of gasoline. That's what it would have taken to move the United States into the kind of energy-research and exploration era that would have set the stage for an eventual reduction of imports. And that price of 65-68¢/gal. does not even consider price elasticity. Taking elasticity into consideration, gasoline prices may have been expected to go up another 5 to 10%.

A question of considerable interest to energy observers at that time was whether deregulated oil and gas prices, combined with President Ford's proposed import fee, would have generated enough funds in the United States to pursue a vigorous energy investment program. As the calculations below reveal, the financial arithmetic of the 1975 State of the Union Address was about right.

1975 price of imported oil, delivered in U.S. $12
Plus long-run import fee . . . . $ 2
Final price under new proposal $14

Price differential to OPEC ceiling (in dollars per barrel):

	1974 prices	Excise tax	New prices	Differential to OPEC price
Old oil	5.25	2.00	7.25	6.75
New oil	11.50	2.00	13.50	0.50

Revenue generated by rise of U.S. prices to OPEC ceiling:

	%	1975 production (MMbpd)	Differential to OPEC price ($/bbl.)	Annual revenue (billions)
Old oil	60	5.04	6.75	$12.4
New oil	40	3.36	0.50	$ 0.6
	Totals	8.4		$13.0

Revenue on excise tax:
Oil, domestic and imported ....$12 billion
Oil and gas ....$20 billion
Funds generated by rise of domestic crude-oil prices to OPEC ceiling, net of excise tax:
Domestic oil ....$13 billion
Oil and gas ....$30 billion

Thus, the increase in oil and equivalent gas prices to the OPEC ceiling, defined here as delivered 1975 OPEC prices plus a $2.00 per barrel import fee, would have been sufficient to cover the proposed excise taxes ($20 billion) and the required windfall profits tax ($10 billion), and in addition it would have generated another $20 billion per year for capital investments. At that point, domestic crude oil would have cost $14 per barrel and the price *increase* of natural gas would have been 76¢ per Mcf.

Most of the immediate actions President Ford proposed in his 1975 State of the Union Address were reasonable enough. If U.S. oil prices had really been allowed to reach the OPEC ceiling, imports would probably have declined on their own. Whether this would have put downward pressure on OPEC prices is uncertain, but if it had not, nothing else would have.

There was one feature in the energy program that could only have led to problems: the fee on imported crude oil and products that was scheduled to reach $3 per barrel on April 1, 1975, and ultimately to be replaced by legislated fees of $2 per barrel. Like other provisions of President Ford's energy plan, this one succumbed to legislative debate. The proposed import fee represented a complete reversal of U.S. policy, since the government had always taken the position that oil prices must be kept low. Now it was proposing to raise the price of imported oil, negating everything it had said previously.

Very briefly, here are the remaining points of the President's 1975 energy program and a discussion of their merits.

Immediate Measures

§ Opening of outer continental shelf and naval petroleum reserves in Alaska—A good move.

§ Clean-air amendments to allow greater coal use—Painful but necessary; a good move.

§ Revision of strip-mining legislation—A good move.

§ Expediting conventional nuclear power program—A good move.

§ One-year investment tax credit extended additional two years for construction of power plants not using gas or oil—Sorely needed; a good move.

§ Selective reform of state utility commission regulation—Not specified, hence no comment.

§ Tariffs, import quotas or price floors to achieve energy independence—Not needed if domestic prices are deregulated.

Long-Term Measures

§ Mandatory thermal efficiency standards on new buildings—Not needed; would have become automatic with unregulated fuel prices.

§ Tax credit up to $150 for home insulation—See comment above.

§ Help low-income families purchase insulation—Cost of the bureaucracy needed for that would probably have exceeded price of insulation.

§ Defer auto pollution standards by five years—Good move, especially since driving would have been cut back through deregulation of fuel prices.

§ The strategic storage program with one billion barrels for domestic needs and 0.3 billion barrels for national defense, and covering a normal U.S. consumption period of approximately 75 days, was a bit puzzling. The blueprint of this policy was straight out of the *Project Independence Report,* where the cost was estimated at $8.2 billion over 10 years (p. 10). The obvious "justification" of this program was that it would reduce U.S. vulnerability to another embargo. But if the oil industry can be revitalized through deregulated prices, this program may not be needed. In fact, Project Independence only considers the storage cost. The logistics of distribution, should an embargo really occur, will dwarf the storage problem.

Of course, most European countries do have a 60- to 90-day storage program, but they have no production of their own to fall back on if the going gets rough. Moreover, these are smaller geographic entities that present fewer distribution problems. Therefore, what is good for them is not necessarily good for us.

Overall, then, the President's 1975 energy program—all things considered—would have helped the United States overcome its energy crisis, primarily because it envisioned the deregulation of oil and gas prices. Without that feature, the industry would have been unable to provide for U.S. energy needs. As was to be expected, and in accordance with a prediction to that effect by the author (*World Oil,* February 15, 1975), Congress did not go along with deregulation beyond a token inflation adjustment of oil prices that became a feature of the December 1975 legislation on energy. Once again, politics had triumphed over economics.

ii. **The Federal Energy Administration**

A bountiful supply of energy is one of the many things that the American people have always taken for granted. In the two decades preceding the oil embargo, U.S. consumption of all forms of energy had more than doubled. The assumption that plentiful supplies of energy would always be available had been seriously challenged as early as 1967, when the energy industries began to broadcast warnings of trouble ahead. It was not until 1971, however, that a few public and political leaders began to perceive the problem. More effective than warnings by industry leaders, the Arab oil embargo brought home to the American people the fact that America was dependent on foreign oil to meet its growing energy demands.

On June 4, 1971, President Nixon delivered a message to Congress which summarized America's energy situation.[1] In his message, the President challenged the Democratic Congress to enact legislation that would permit the Administration to deal with a forecasted energy shortage. Facts showed that prior to 1967 the demand for energy grew at about the same rate as the GNP. However, from 1967 to 1971, energy consumption had grown at a faster pace, and forecasts of energy demands in the decade ahead were revised upward significantly. Relatively low-cost energy was cited as part of the reason for the rapid growth in demand. Energy had been an attractive bargain in this country, due mainly to artificially controlled low prices, and demand had responded accordingly. President Nixon outlined two broad goals in his message to Congress: legislative action to create incentives for supplying adequate energy and protection of the environment in the decades ahead.

President Nixon's proposed energy program included the following provisions: first, the establishment of a research program to create clean energy through nuclear reactors and to provide support for smokestack clean-up and expanded outlays for the gasification of coal; second, a leasing program to make available the energy resources on federal lands—areas specifically cited were the outer continental shelf, oil shale lands, and geothermal sites; third, an expansion of the U.S. uranium-enrichment capacity; fourth, a revision of housing standards designed to conserve energy; and fifth, a federal energy authority within a Department of Natural Resources to unify all energy resource development programs:

> The nation has been without an integrated energy policy in the past. One reason for this situation is that energy responsibilities are fragmented among several agencies ... authority is divided, ... responsibility for considering the impact of one energy source on another

is not assigned, . . . there is no single agency responsible for developing new energy sources. If the government is to perform adequately in the energy field, then it must act through an agency which has sufficient strength and breadth of responsibility.

On August 15, 1971, Phase I of President Nixon's Economic Stabilization Program was put into effect, which envisioned the freezing of all prices, including prices of crude oil and petroleum products, for 90 days. Phase II of the program lasted from November 14, 1971, until January 10, 1973. During this time all crude oil and petroleum products were traded under the overall price rules of this program. Beginning with Phase III, on January 10, 1973 the oil industry was returned to a free enterprise structure with instructions "to exercise voluntary restraints on price increases." Two months later, on March 13, 1973, mandatory controls were re-imposed on the oil industry.

Recognizing the gravity of the overall energy situation, President Nixon delivered a second energy message to Congress on April 18, 1973. Here is an excerpt of his opening remarks[2]:

At home and abroad, America is in a time of transition. Old problems are yielding to new initiatives, but in their place new problems are arising which once again challenge our ingenuity and require vigorous action. Nowhere is this more clearly true than in the field of energy.

As America has become more prosperous and more heavily industrialized, our demand for energy has soared. Today with six percent of the world's population, we consume almost a third of all the energy used in the world. Our energy demands have grown so rapidly that they now outstrip our available supplies, and at our present rate of growth, our energy needs a dozen years from now will be nearly double what they were in 1970.

In the year immediately ahead, we must face up to the possibility of occasional energy shortages and some increase in energy prices.

In this second energy message, some 21 months after the first one, President Nixon outlined a comprehensive program to provide for the nation's current and future energy needs. This program called for an increase in domestic fuel production to minimize risks to the national security due to supply interruption. A balance between national security considerations and continued protection of the environment at reasonable energy prices was stressed.

The main elements of President Nixon's energy plan of April 18, 1973, included:

i. A proposal to amend the Natural Gas Act so that competitive forces of the market system would determine the prices paid to producers by interstate pipelines for new domestic natural gas, thus removing the price-control function from the Federal Power Commission. The Natural Gas

Act of 1938 had been originally enacted to give authority to the FPC to regulate the transportation, sale and resale of natural gas by interstate pipelines. Although the Act specifically precluded the regulation of the production of natural gas, the Supreme Court, in the landmark Phillips case of 1954, held that the Natural Gas Act applied to sales by producers in interstate commerce. Congress twice passed legislation to effectively deregulate natural gas at the well head, once in 1950 and once in 1956, each time only to be vetoed by Presidents Truman and Eisenhower, respectively.

ii. A stepped-up program for leasing the outer continental shelf, particularly in the new frontier area of water depths beyond 200 meters, and in other specific areas. A program of surveillance of pollution, including aerial inspection, coordinated with the anticipated increase in offshore drilling was also announced.

iii. An urgent request that Congress act with haste on the Alaskan Pipeline Bill which had been pending since May 1972.

iv. An announcement that the President had ordered the preparation of an environmental impact statement in anticipation of an oil shale leasing program.

v. An urgent request for adoption of various policies relative to coal production and use, to geothermal energy, and to nuclear power.

vi. A restructuring of the Mandatory Oil Import Program to lower import costs by removing tariffs and by limiting license fees on imported oil to volumes exceeding the 1973 level. The President also announced specific provisions for stimulating the construction of domestic refineries and plans to provide for increased storage of crude oil in order to minimize the impact of possible future supply interruptions.

vii. Specific proposals touching on deepwater ports, conservation, research and development, and international cooperation.

viii. Re-proposed legislation for the creation of a federal energy agency to develop and coordinate important energy policy functions and programs. For the interim, President Nixon announced the establishment of a Special Energy Committee within the Office of the President and an Oil Policy Committee in the Treasury Department, the latter having total responsibility for oil imports.

Mid-1973 saw a marked change in the attitude of the Democratic Congress in the area of energy. U.S. political support for Israel was causing problems with the Arab nations, the world's largest exporters of crude oil. Also, due to the sustained clamor of oil industry leaders, the general public and consequently their political representatives were slowly beginning to realize that domestic oil production had peaked and was on the decline.

On June 29, 1973, President Nixon announced the creation of the Energy Policy Office (EPO) replacing the Special Energy Committee.[3] This office would be responsible for the formulation and coordination of all energy policies. Renewed appeals were made to Congress to pass legislation creating a cabinet-level Department of Energy and Natural Resources (DENR), and an Energy Research and Development Administration (ERDA). In addition, the President asked the American people to reduce personal consumption of energy by 5% over the next 12 months, and he directed each executive department and agency to participate in a government-wide program to reduce energy consumption by 7%. He called for a reduction in highway speed and sent state governors a personal letter urging them to use the state legislatures to cooperate in a program of energy conservation.

The outbreak of fighting between Israel and the Arab nations in October, 1973, led to an Arab oil embargo of the United States and certain other countries in Europe. For the United States, the embargo meant a reduction in oil imports of at most 1.5 million barrels of oil and products per day during the period December 1973 through April 1974, and perhaps considerably less if reduced inventory draw-downs are taken into consideration. In response to this "crisis," President Nixon addressed the nation on November 7, 1973, and delivered essentially the same message to Congress the next day. In these messages, the President renewed his request for legislation which would create a cabinet-level department responsible for the coordination of U.S. energy policies. He further called for the competitive pricing of natural gas. In addition, President Nixon's emergency program dealt with the Alaska Pipeline, and it called for the implementation of reasonable standards for the surface-mining of coal, a simplified procedure for siting and approving of electric energy facilities and the approval for construction of deepwater ports. Presidential authority to restrict the consumption of energy in the public and private sectors, reduced highway speeds, waivers on some air and water quality laws, increased oil production from Naval Petroleum Reserve #1 (Elk Hills, Calif.), and the initiation of daylight savings time were also requested.[4] In closing, the President stated:

> Because of the critical role which energy research and development will play in meeting our future energy needs, I am requesting the Congress to give priority attention to the creation of an Energy Research Agency separate from my proposal to create a Department of Energy and Natural Resources. This new agency would direct a $10 billion program aimed at achieving a national capacity for energy self-sufficiency by 1980. This new effort to achieve self-sufficiency in energy, to be known as Project Independence. . . .

On November 19, 1973, the Congress passed the National Energy Emergency Act which (a) restricted gasoline sales on Sunday, (b) set the maximum highway speed at 55 mph, (c) banned ornamental lights on commercial establishments, (d) reduced fuel allocations for general aviation, and (e) banned Christmas lights on residences.

The National Energy Emergency Act was followed, a week later, by the Emergency Petroleum Allocation Act, EPAA (P.L. 92-159). The act vested the executive branch with the power of mandatory allocation of all petroleum products and with the petroleum price control authority formerly under the jurisdiction of the Cost of Living Council. The act is characterized by generalized regulations creating price controls at two points in the chain of commerce from the well head to the consumer. At the lease line, the EPAA adopted the concept of two-tier crude oil pricing. Oil produced from properties that were on production in 1972 was defined as "old" oil and assigned a price consisting of the posted price on May 15, 1973, plus $1.35/barrel. Exempted from the old oil price was stripper oil from wells producing less than 10 barrels of oil per day. To prevent early plugging and abandonment of wells, stripper oil remained unregulated under the EPAA.

To provide incentives for workovers and new exploratory drilling, so-called new and released oil was also exempted from price controls. "Released" oil was defined as production from old properties exceeding the base production control level (BPCL), i.e., exceeding that volume of production which these properties produced during corresponding months in 1972. "New" oil was defined as oil from newly developed properties not under production in 1972.

At the consumer level, the Emergency Petroleum Allocation Act permitted product prices to be increased by a cost pass-through system. Since the two-tier pricing of crude oil at the production end, coupled with a cost pass-through system at the refinery level, was expected to lead to "inequitable" product prices at different locations in the United States, the Emergency Petroleum Allocation Act envisioned the establishment of the so-called entitlements program. This program, which became operative in November of 1974, attempted to equalize multi-tiered crude oil prices at the refining level.

The Birth of FEA

In December, 1973, the Federal Energy Office (FEO) was created by executive order. The FEO took over the responsibilities of the Energy Policy Office (EPO) and the Office of Petroleum Allocation (OPA). The FEO also assumed new powers which had been authorized by the Emergency Petroleum Allocation Act on November 27, 1973. The Federal Energy Office was considered only an embryonic office that would ultimately be replaced by a more encompassing Federal Energy Administration.

After much debate, the U.S. Congress passed P.L. 93-275, creating the Federal Energy Administration. The legislation was signed into law on May 7, 1974, and the Act was officially implemented on June 27, 1974. The Act creating the FEA was initially scheduled to expire on June 30, 1976. However, President Ford asked for an extension of the FEA, after Congress expressed its wish to continue the petroleum allocation and price-control program. Before Congress could act on its extension, the FEA's term expired and the entire allocation and price-control machinery reverted to the jurisdiction of the FEO. After a 15-day hiatus in its continuity, the FEA obtained a new release on life when P.L. 94-385 was passed and signed by President Ford.

The Federal Energy Administration presently has jurisdiction over the following areas:

1. Fuel allocation and petroleum pricing regulations
2. Energy information and analysis
3. Energy planning for making the nation less dependent upon foreign sources of oil
4. Energy conservation and environment
5. Strategic petroleum reserves
6. International energy affairs

The top level of the FEA organization is comprised of an Administrator, two Deputy Administrators, a General Counsel and six Assistant Administrators, one for each of the above-named functions.

The Energy Policy and Conservation Act of 1975

The second wave of regulations came with the Energy Policy and Conservation Act, EPCA, (P.L. 94-163), signed into law on December 22, 1975. The EPCA amended and extended the EPAA, mainly regarding controls on the crude oil producer. The main issue addressed by the EPCA was an amendment of the two-tier pricing system. However, other issues such as property definition, price differentials due to gravity, and incentives for enhanced oil recovery were also included. The primary crude oil provisions of the Energy Policy and Conservation Act:

—placed all crude oil under rigid two-tier price controls;

—rolled back the price of upper-tier oil, formerly new, released, and stripper oil;

—redefined the base production control level (BPCL) for lower-tier oil, formerly old oil;

—established a mechanism for the escalation of crude oil prices over a period of 40 months; and

—mandated the strategic petroleum reserves program.

The pricing and strategic reserves provisions are discussed in more detail in the following paragraphs.

Pricing. The new pricing policy under EPCA made all domestic crude oil subject to a composite price ceiling of $7.66 per barrel, effective February 1, 1976. This initial price was based on a lower-tier price of $5.25 per barrel and an upper-tier price of $11.28 per barrel. In addition, EPCA assumed lower-tier oil to represent 60% and upper-tier oil 40% of the total domestic crude oil production.

A monthly upward adjustment of the composite price of $7.66 per barrel was provided for under the terms of the Energy Policy and Conservation Act in order to adjust for inflation, as defined by the GNP-deflator, and to provide for a production incentive of not more than 3%. The two adjustments combined were not to exceed 10% per year.

Other pricing provisions of the Energy Policy and Conservation Act allowed the mandatory control program to convert automatically to a discretionary program at the end of 40 months. The Act also provided that Alaskan oil could be excluded from the composite crude oil price after April 1977, and it envisioned a review of current regulations with the goal of partly dismantling the price-control program. These and other prerogatives were subject to disapproval by either house of Congress for up to 15 days after they had been put into effect.

Strategic Petroleum Reserves. The Strategic Petroleum Reserves program mandated that crude oil storage of 150 million barrels of petroleum be created within three years, with increases to 400 million barrels in seven years. It was anticipated that up to one billion barrels of crude oil would ultimately exist in emergency storage at strategic locations throughout the United States. This volume of crude oil would provide approximately three months of emergency supply in the event of another crude oil embargo.

Other provisions of the Energy Policy and Conservation Act are:

i. Standby energy emergency authority allowing the President to deal with severe energy emergencies as they may arise in the future;

ii. International energy authority allowing the United States to participate in a variety of International Energy programs;

iii. Coal conversion authority to permit the conversion of oil- and gas-fired utility and industrial boilers to coal;

iv. Appliance labeling and automobile efficiency standards.

The Entitlements Program

As mentioned earlier, the entitlements program was broadly defined in the Emergency Petroleum Allocation Act of November 1973. The program became operative a year later, and it was amended in the Energy Policy and Conservation Act of December 1975.

Based upon the assumption that is is unfair to allow some refiners to have access to more lower-tier crude oil than others—even those who developed and produced the oil—the FEA has made rulings to essentially equalize the cost of crude oil feedstock to all refiners. Under the entitlements rules, the FEA determines the ratio of upper- to lower-tier crude oil available in the nation. A refiner who processes a percentage of upper-tier crude oil higher than this ratio is issued entitlements; refiners who process a lower percentage of upper-tier crude oil must buy these entitlements at a price set by the FEA. In effect, the entitlements program causes a transfer of the value of oil rather than the physical transfer of the oil itself. As it works out, the entitlements program rewards the independent refiner located in non-oil-producing areas at the expense of the integrated oil companies. No regard is given to the tremendous investments made by the integrated oil companies in finding, developing, and producing domestic crude oil, nor to the future investments that are required to maintain production. On the whole, the entitlements program reverses the multi-tiered oil price structure (lower, upper, stripper and imported) at the refining level.

An impartial observer looking at the entire crude-oil pricing system can hardly escape the conclusion that the federal government is slowly destroying free enterprise in the oil industry, replacing it with a compilation of rules upon rules to satisfy special interest groups and politically powerful constituents mainly in the non-oil-producing states. Recounting an actual incident resulting from the entitlements program will support this contention.

As stated earlier, the entitlements program was conceived to reverse, at the refining level, the multi-tiered price-control structure that is imposed on the crude oil producer. However, when the program went into operation, it did not and could not include all the adjustments which are made automatically by the activities of the free market system. A serious problem developed with respect to the East Coast residual fuel market, where Amerada Hess qualified for entitlements while its competitors, whose refineries are located abroad, did not. All five of the area residual fuel suppliers, consisting of Amerada Hess, Asiatic, Exxon, NEPCO and Texaco, had refineries in the Caribbean area, but only Amerada Hess had its refinery on U.S. territory (in the Virgin Islands). Obviously, this problem was brought about by fundamentally inadequate and imperfect government regulation, since each company based its choice of market area and related refinery location on optimizing its circumstances in a free market as it existed prior to EPCA. Not surprisingly, the government's solution was more regulation.

NEPCO's declining competitive situation, relative to Amerada Hess, led the FEA to award the importer a number of entitlements on a monthly basis under the allocation exception procedure. Through this adjustment, all of NEPCO's competitors in effect pumped money into NEPCO in order to temporarily stabilize the situation. Soon thereafter, the FEA required the restructuring of the entire oil import fee program. Again, due to its refinery location on U.S. territory in the Virgin Islands, Amerada Hess, through windfall circumstances, found itself in a position whereby it could have as much as a $3.00 per barrel advantage over its four competitors.[5]

To offset this second competitive inequity, the FEA reduced by one-half the number of entitlements that would accrue to all domestic East Coast refiners of residual fuel, this time including the Virgin Islands. Also, entitlements were cancelled on refinery capacities less than 5,000 barrels per day. As an offset, importers of residual fuel would receive 30% of the entitlement value of a barrel of crude oil for each barrel of residual fuel that they imported into the East Coast. No such privileges were extended to other parts of the nation. These new FEA rules were believed to reduce the competitive advantage of Amerada Hess from $3.00 per barrel to only 60¢ per barrel. This is not too bad for rule-making, but it is no substitute for the equilibrium which the free market automatically provides.

The entitlements program has created another ironic situation for the integrated oil companies. Over the years, these companies have made vast investments in capital and manpower along the entire supply chain, from exploration for oil and gas to sales at the pump, for the sole purpose of having assured access to their own supplies of crude oil. Since the allocation program is premised on equality of distribution regardless of ownership, the FEA has established a "buy/sell" program requiring crude oil "haves" to sell a portion of their supplies to their competitors (the crude oil "have-nots," who have made no such investments) at a price set by the FEA at a level below the replacement cost of crude oil. Under the allocation and entitlements rules, approximately 400 million barrels of oil per year, or about 15% of the U.S. production, are subject to forced sale by the following 15 integrated U.S. oil companies: AMOCO, ARCO, Citgo, Continental, Exxon, Getty, Gulf, Marathon, Mobil, Phillips, Shell, Socal, Sun, Texaco and Union.[6]

Prior to the commencement of any allocation quarter, the FEA publishes a buy/sell list containing (a) the quantity of crude oil that each refiner-buyer may purchase, (b) the total allocation obligation of all refiner-sellers, (c) the fixed percentage share for each refiner-seller, and (d) the quantity that each refiner-seller will be obligated to offer for sale during that allocation quarter. Sales transactions must take place within 15 days of the publication of the FEA buy/sell list. The FEA is notified if the desired volume of transactions does not occur in the prescribed time. If a small refiner is unsuccessful in selling entitlements within 30 days, one of the 15 integrated oil companies is directed by the FEA to purchase them. The net effect of the allocation-entitlement program is the direct subsidy of one's own competition in a fashion completely contrary to the free market system.

To the extent that the have-not refiners do not always exhaust their option to buy domestic crude oil, the question arose as to how to handle the carry-over of unsold oil quotas from previous quarters. Beginning with the quarter September-November 1975, a distinction was made between "primary" and "secondary" sales obligations. The carry-over of unsold oil became a primary sales obligation in any new quarter. Since the accumulation of unsold obligations could landslide and cause severe problems for one of the 15 integrated companies, the primary sales obligations are kept to a minimum. No refiner-seller is required to sell any of its secondary obligations (current quarter crude sales) until all 15 of the integrated companies have sold 80% of their primary obligations, the accumulated unsold crude sales obligations from previous quarters.

Thus there can be no doubt at all: government regulations are slowly and methodically strangling the free enterprise system by removing incentives to produce and by rewarding businesses that otherwise could not exist.

The Inherent Contradiction in the Energy Policy and Conservation Act

The United States' apparent inability to implement meaningful steps toward reducing its energy shortage is due primarily to regulation, but not necessarily to the regulator. While the Federal Energy Administration has lent support to the view that higher energy prices are an indispensable prerequisite to solving the shortage, that agency was nevertheless saddled with a mandatory pricing policy through various Congressional mandates.

The Energy Policy and Conservation Act provides a clear example of the FEA's regulatory dilemma. As was pointed out, the original version of EPCA mandated a pricing policy that provided for a 7% maximum inflation adjustment in the composite price of crude oil, plus a 3% maximum production incentive. Nothing could be simpler, it would seem, than the implementation of such a clear-cut mandate. Let us take a look at just this original 7%-3% pricing provision, in lieu of a full discussion of the subsequent events that led, among other things, to the elimination of the 7-3 rule and its replacement by a flat 10% regulation. A discussion of the sticky statistical problem of determining the actual upper- and lower-tier prices or the actual volumes or percentages of upper- and lower-tier oil will also be omitted. These are administrative problems that do not concern us here. Let us instead focus on the conceptual problems inherent in EPCA's pricing provisions.

Let us suppose, then, that lower-tier crude oil production was exactly 60% of domestic production in February 1976 and that $5.25 per barrel of lower-tier oil is a correct price. Given the mandated composite price of $7.66 per barrel, that established the upper-tier crude oil price at $11.28 per barrel, leading to the well-known price roll-back of new and released oil by about $1.18 per barrel.

Of course, the U.S. economy behaved somewhat abnormally in 1975/1976 by exhibiting a rate of inflation only moderately above normal, but we will simplify matters here by assuming that the GNP-deflator returns to a 7%-plus rate and therefore ceases to confound the oil-pricing issues.

Under those circumstances it is clear that the intent of the 7% rule was to keep the real prices of both upper- and lower-tier oil from declining. Quite apart from the fact that the GNP-deflator must necessarily understate the inflation rate in the hottest industry of an otherwise so-so economy, EPCA, by placing a maximum ceiling on the inflation adjustment, in effect conveys this message to the oil producers:

You will be rewarded for *our* success in controlling inflation, and *you* will be punished for *our* failure to do so.

Let us clarify the issue even more by introducing the heroic assumption that the GNP-deflator does indeed run at 7% per year over the 40-month period of mandated composite crude oil prices. In that case, both the upper- and the lower-tier prices would have to be raised by 7% per year just to keep the real price constant. That leaves the problem of how to apply the 3% annual production incentive. This can be done in an infinite number of ways, but three rather obvious possibilities warrant scrutiny.

1. On the premise that it might be desirable to eliminate the two-tier price differential, the incentive might go exclusively to lower-tier crude oil. This would raise the incentive on lower-tier crude oil to something like 13.3%, enough to raise the price of lower-tier crude oil sufficiently, given the lower base, to perpetuate the price differential. For example, the February 1976 differential was $6.03 per barrel (from $11.28 − $5.25); under the assumed condition of exclusive application of the incentive to the lower-tier price, the differential five years later would still be $6.03 per barrel. However, the resulting perpetuation of the price differential is in conflict with the stated objective of ultimately eliminating such a differential. Hence, the exclusive allocation of the production incentive to lower-tier oil cannot achieve its objective and the policy becomes futile.

2. If the 3% price differential is applied equally to the lower- and upper-tier prices of crude oil, then that differential would rise with time and the distortion of current prices would be aggravated. For example, the initial price differential would rise from $6.03 per barrel from year 1 to $9.70 per barrel in year 5 and to $15.64 per barrel in year 10. Such a substantial price differential would set up great incentives to increase upper-tier oil production, both legally through exploration, through exotic secondary or tertiary recovery methods, and the like, and not so legally through carefully timed non-fracture treatments or non-acid stimulations, to name two such devices that would speed a well's life-cycle from the lower-tier stage to stripper well status.

3. If the price differential becomes prohibitive with time under an equal application of the 3% production incentive, it becomes doubly prohibitive if that incentive is applied collectively to upper-tier oil, so we will not waste our time discussing this case. Instead, let us direct our attention to another feature of EPCA.

Suppose the Energy Policy and Conservation Act succeeds in stimulating the production of new crude oil to such an extent that 1977 finds us with the same volume of domestic production, except that the production of lower-tier oil has declined by half a million barrels per day (from 4.9 to 4.4 million barrels), a 10.2% decline that is not unreasonable. This means, of course, that the upper-tier production would have risen from 3.3 to 3.8 million barrels per day. Such an event would shift the price base of lower-tier oil from 60% (4.9/8.2) to 54% (4.4/8.2).

Since the mandated composite crude oil price in the twelfth month is $8.43 per barrel, the shift in weights would prevent the application of the full 10% price increase on upper- and lower-tier oil. At 10%, the composite price would be

$$1.10 \times (.46 \times 11.28) + 1.10 \times (.54 \times 5.25) = \$8.88/bbl.$$

That is 40¢ per barrel above the mandated ceiling, so the full application of the 10% price increase becomes illegal.

Suppose it is decided to apply only the 7% inflation factor, leaving out the 3% incentive? This would result in a price of:

$$1.07 \times (.46 \times 11.28) + 1.07 \times (.54 \times 5.25) = \$8.59/bbl.$$

This is still above the mandated price ceiling, by 16¢ per barrel. The law is the law, so the FEA will have to whittle away on the inflation adjustment. By how much?

$$\$8.43 = X \times (.46 \times 11.28) + X \times (.54 \times 5.25)$$
$$X = 1.05, \text{ or little more than } 1\%!$$

The total price increase is only 5% due to the change in mix of upper and lower-tier oil instead of the sum of 7% for inflation and 3% for incentive specified by the law. That is, the law is designed in such a way as to discourage domestic production.

The Energy Conservation and Production Act of 1976

On August 15, 1976, the EPAA and the EPCA were amended by the Energy Conservation and Production Act, ECPA, (P.L. 94-385). The major provisions of the ECPA were the deregulation of the price of stripper oil and the establishment of the escalation factor for the composite price of crude oil at 10%, thus freeing it from the GNP-deflator. This Act also cleared up confusion in the definition of a producing property. All these new provisions became effective on September 1, 1976. Under the Energy Conservation and Production Act, the price of stripper production, which comprised about 14% of U.S. domestic production, advanced to meet the cost of imported oil of similar quality laid down in the United States. Posted prices for stripper production averaged about $14.00/barrel in the period September-December 1976, with a spread of about 50 cents each way depending on quality. Also, the new definition of producing property allowed approximately 2.3% of lower-tier oil to be reclassified as stripper oil, to bring that category up to 16.3% of total domestic production.

As it turns out, the original assumption of a 60-40% distribution of lower- and upper-tier oil was in error, and the news broke in November of 1976 that the average U.S. crude oil price had been considerably higher than the price mandated by EPCA/ECPA. This honest FEA mistake required either a legislative upward revision of the pricing provisions under the existing acts or a roll-back of crude oil prices. Economic criteria would have strongly favored a legislative upward revision of the composite crude oil price, while political sentiment at that time leaned towards price roll-backs. Not surprisingly, the outcome of this economic/political conflict was a price roll-back in the amount of 20 cents per barrel of upper-tier oil, effective January 1, 1977.

Since the creation of the Federal Energy Administration, U.S. crude oil production has continued to decline while crude oil imports have risen at accelerating rates. Various conservation programs caused a temporary reduction in energy consumption in 1974 and 1975; however, recent statistics indicate that this involved no more than a one-time shift in consumption patterns. Already, the pre-embargo energy consumption rate appears to have resumed in 1976.

The debate over energy regulation has served to bring two opposing views into sharp focus. One view, held by proponents of central planning, places total faith and reliance in bureaucratic efficiency and knowledge. The second, the free market view, holds that the natural adjustment forces at work in a freely competitive market will lead to an optimum solution for the nation and the world. Unfortunately, the free market philosophy is becoming increasingly suspect in the United States. For better or for worse (and most likely for worse), the implied prospects in this country are for more centralization and tighter control and, therefore, for greater inefficiency in the U.S. oil industry.

References

1. President Nixon's Energy Message to Congress, June 4, 1971, *Weekly Compilation of President Documents,* Vol. 7, No. 23, June 7, 1971, pp. 855-866.
2. President Nixon's Energy Message to Congress, April 18, 1973, *Weekly Compilation of Presidential Documents,* Vol. 9, N. 16, April 23, 1973, p. 389-406.
3. President's Statement on Energy Summary Outline—Fact Sheet, Office of the White House Press Secretary, June 29, 1973.
4. Address by the President on the Energy Crisis, *Commerce Clearing House, Inc., Energy Management,* Vol. 1 Sections 549 through 566, pp. 555-570.
5. "Crude Oil Allocation and Refinery Yield Control," *Commerce Clearing House, Inc. Energy Management,* Vol. 1, Sec. 3524, Entitlements, pp. 3354-3356.
6. "Crude Oil Allocation and Refinery Yield Control," *Commerce Clearing House, Inc. Energy Management,* Vol. 1, Sec. 3523 Buy/Sell program, pp. 3353-3354.

9
Energy and The Overall Economy

i. Oil and the U.S. Economy:
A View from the Arab Side

The Ninth Arab Petroleum Congress was held in Dubai, United Arab Emirates, March 10-16, 1975. Sponsored by the League of Arab States, this was the first meeting of its kind to be held at a time when Arab control over Arab oil and gas was virtually complete. For the sponsors, this occasion had great significance.

Since economic policies are not made on the basis of facts, but instead on how political leaders perceive and interpret those facts, the Ninth Arab Petroleum Congress provided an excellent opportunity to assess Arab perceptions of the political-economic environment of oil-importing nations, particularly that of the United States. This section contains an analysis of these Arab perceptions, on the basis of speeches and papers delivered at the Ninth Arab Petroleum Congress.

The Congress was divided into three sections, with separate papers—economics, processing, and production. The following remarks will be limited to the section on economics.

By and large, papers were well thought through and worthy of presentation at an international conference of this kind. Many Arab authors understood the economic dilemma that labor unions pose to industrialized countries, a problem that still goes largely undetected in the affected nations.

Possibly the weakest discussions by Arab authors stemmed from a misunderstanding of what makes the U.S. economy tick. While it is not very difficult to agree with the notion that there is room for improvement in the U.S. economy, our criticism (as opposed to Arab criticism) of the U.S. economy is based on entirely different grounds.

Arab misconceptions concerning the U.S. economy represent a gap that needs plugging. The United States and the Arab oil-exporting countries are now entering a phase of cooperation in the sense that the two, perhaps for the first time in history, are true and equal partners.

Of course, the fundamental nature of this relation is conflict—economic conflict. The United States' interest is in reduced crude oil prices; it is not likely that the Arab oil-exporting nations will cooperate by reducing them. The Arab oil exporters on the other hand wish to keep *real* crude oil prices at least constant. At current inflation rates in the United States and elsewhere, this implies rising nominal crude oil prices. It's just as unlikely that the United States will cooperate in achieving this Arab objective.

Unfortunately, this economic conflict has the potential of degenerating into armed conflict. Defense Secretary J.R. Schlesinger, interviewed on the Public Broadcasting Services program *Straight Talk,* Monday, March 31, 1975, had this to say when asked if the United States was prepared to endure another oil boycott: "We would not expect readily to tolerate such a renewed boycott . . . the reaction of the United States would be far more severe this time than last time. . . ."

U.S. citizens might resent this attitude. The issue here is not the question of world peace, as it was, or was thought to be, in Vietnam—the now discarded domino theory. The issue is a wealth transfer of no more than 2% from the United States to the oil-exporting nations. Are we really prepared to resort to warfare for that?

Let us hope that the conflict will remain only an economic conflict, one that will be resolved through negotiation. And for negotiations to be useful, it is vitally important that the two partners know each other's position. This is especially true now that equality in partnership has been attained.

OPEC countries in general, and Arab oil-exporting countries in particular, need to familiarize themselves with the economic realities of the United States and other oil-importing nations. This is the crucial challenge facing OPEC nations today.

One opinion commonly held by Arab authors is that the U.S. government and the U.S. oil industry work hand in hand. Nothing could be further from the truth. Actually, the U.S. government is blatantly antagonistic toward its own oil industry, especially on the legislative level.

President Ford's energy plan, announced in his 1975 State of the Union Address, preceded the Ninth Arab Petroleum Congress by about one month. Suffice it to point out here that some provisions of that plan made good economic sense, notably the proposed decontrol of domestic oil and new gas prices. Other provisions made good political sense, such as the proposal to help low-income families purchase home insulation. And, of course, many provisions were nonsense: tariffs, import quotas and price floors, as examples.

Still, despite many imperfections, President Ford's energy plan would have worked. But it had absolutely no chance of surviving the legislative phase of its enactment, certainly not with the belligerently anticorporate and anti-oil 94th U.S. Congress. This was pointed out to the delegates of the Ninth Arab Petroleum Congress, and examples were given of legislative energy proposals that had at one time or another been considered by the 93rd and 94th U.S. Congresses. These were, nearly without exception, detrimental to the U.S. oil industry.

Former Saudi Arabian Oil Minister Abdullah Tariki told delegates that what the Arab oil-producing countries need are more technicians and fewer politicians. This surprising observation must highlight a common problem— for it certainly applies to the United States as well.

Another Arab author suggested that the trade union movement and public opinion in the industrial countries have become so powerful as to resist a rate of unemployment exceeding 5%, and they may even press for reducing it to 3%. To take issue with the misconception would require another book, since this involves the entire complex issue of the stagflation presently facing all industrialized countries. As was pointed out in Chapter 3, the increase in crude oil prices had very little to do with this stagflation. Still, in response to the 5 or 3% unemployment argument, the U.S. unemployment rate is upward of 7%, very much to our government's obvious and visible frustration.

Another Arab belief, also shared by many other countries, is that the U.S. government (in cooperation with the U.S. oil industry) deliberately pursued a policy of higher crude oil prices to:

1. Inflate U.S. oil company profits and induce them to develop alternative energy sources;

2. Make U.S. industries more competitive vis-à-vis Europe and Japan, who have to import virtually all of their crude oil needs.

This theory ignores the fundamental issue that underlies OPEC's motives and market structures: the transfer of real wealth from oil-importing to oil-exporting countries. All other factors, such as profit rates in given industries or changing exchange rates of world currencies, are mere surface phenomena. Once this is understood, it becomes clear that the United States would have been much better off purchasing Arab oil at $3.00 per barrel, if it had been able to do so.

If the U.S. government had really been concerned about future energy sources, it would have made much more sense to buy Arab oil at $3 per barrel rather than to push the price to the present level of $11, as it was alleged to have done. The resulting savings could have subsidized development of alternative U.S. energy sources. This is illustrated more clearly in Table 9-1.

As can be seen in Table 9-1, and with assumption of a complete reversion of OPEC oil to OPEC control in 1974, the total cost of imported oil over the 12-year period 1974-1985 would have been $122.4 billion at $3 per barrel, as opposed to $231 billion at $11 per barrel. Savings to the United States over this period would have been $108.6 billion.

This is a very conservative estimate, since future oil projections, based on the *Project Independence Report*, anticipated a decline in imports at the higher price. The fact is that 1975 imports were indeed 2.2 billion barrels of oil as shown, but the first quarter of 1976 exhibited an 18% increase over the preceding year's import rate, from 2.2 to 2.6 billion barrels. There is no reason to believe that imports will subsequently decline or even remain at the 2.2 billion level.

The U.S. proposal concerning establishment of a $7-$8 floor on the price of crude oil is often cited as "evidence" that the United States is interested in high crude oil prices. Yet, there is a difference between pushing the price to $11 per barrel as opposed to not wanting it to drop below $8 after it has been pushed to $11 by someone else.

Table 9-1
Why the United States Would Have Benefited More from continued Low-Priced Imports

Year	Projected Annual Oil Imports (billions of barrels)		Payments to Oil-Exporting Nations		Saving Due to Low Oil Prices, ($ billion)
	$3/bbl.	$11/bbl.	At $3/bbl. ($ billion)	At $11/bbl. ($ billion)	
1974	2.3	2.3	6.9	25.3	18.4
1975	2.5	2.2	7.5	24.2	16.7
1976	2.7	2.1	8.1	23.1	15.0
1977	2.9	2.0	8.7	22.0	13.3
1978	3.1	1.9	9.3	20.9	11.6
1979	3.3	1.8	9.9	19.8	9.9
1980	3.5	1.7	10.5	18.7	8.2
1981	3.7	1.6	11.1	17.6	6.5
1982	3.9	1.5	11.7	16.5	4.8
1983	4.1	1.4	12.3	15.4	3.1
1984	4.3	1.3	12.9	14.3	1.4
1985	4.5	1.2	13.5	13.2	0.3
			122.4	231.0	108.6

Projected annual import volumes from *Project Independence Report*, page 31. Interpolation of inbetween years added.

What is involved is the principle of the destruction of capital by reducing oil prices. This problem was discussed in Chapter 2, section *iii*. Still, it is not at all clear that the price floor plan is a particularly effective method for dealing with the problem.

Others agree with this negative assessment of the price floor plan. An official representative of a friendly European government said unofficially that his government had no faith in the plan. Using European finesse, and not wanting to openly oppose the United States on the issue, they actively supported the plan except that they insisted on a price floor so low as to make the whole plan totally ineffective. That was a clever move, for what might have been a battle over principles was thus contained to a disagreement on implementation. The Americans could not understand how anyone could lack the intelligence to see the need for a higher floor, when in fact they had been outmaneuvered.

The last Arab misconception to be discussed here deals with the Arab brain drain, i.e., the loss of Arab talent to industrialized nations, including the United States. The problem in itself is real enough, but it is often stated in terms of Arab talent being exploited by the industrialized or—depending on the speaker's point of view—"capitalist" nations. This is pure rhetoric. The truth is that many Arabs prefer the economic advantages outside their home countries. They like being exploited by capitalists.

A newly graduated petroleum engineer with a U.S. Bachelor of Science degree may receive a monthly salary of $1,300 or more in the United States, and in many other locations worldwide. This is an attractive opportunity. The issue of whether the U.S. technical degree is superior to the Arab degree is totally irrelevant here. The U.S.-educated Arab often has career opportunities (at least in the short run) greatly exceeding those of his Arab-educated brother. Conversely, the Arab-educated engineer or manager is right in complaining that the U.S. education may lead to a salary three times as high as his own.

The basic problem is a talent shortage in many Arab countries. To relieve that shortage, salaries have to be boosted, rapidly and massively. There is no other way. Many Arab countries will simply have to free their minds from the old salary constraints that once were compatible with their economies. They now have the funds for investment, and the best investment they can possibly make is in the formation and retention of their minds.

One alternative is to hire foreign consultants to guide development. These firms often provide local expertise by employing U.S.-trained Arabs. Thus, the Arab governments actually pay the demanded salaries, but the payment goes abroad where the so-called multiplier effects benefit non-Arab countries.

Of course, there is a difference between paying a handful of Arab consultants U.S.-type salaries and paying the entire work force similar or slightly lower salaries. Yet the funds are there, and, additionally, the Arab salaries provide a source of revenue to a whole range of secondary activities. The resultant demands for products might well create new markets sufficient in size to encourage a switch from current importation to domestic production. This would create additional jobs in new industries and act as a catalyst in the transfer of technological know-how. Looked at this way, the story of exploitation of Arab talent becomes one of opportunities missed by Arab countries.

ii. Perceptions and Misconceptions Held by the U.S. Public

The most puzzling aspect of the U.S. energy situation is that it is easy to explain, that economists have seen it in the making for a long time, that policy solutions are obvious—yet nothing concrete is being done to solve the problem.

In 1972, the last full calendar year prior to the Arab oil embargo, the United States consumed petroleum and petroleum products at the rate of 16.2 million bpd, and we imported some 27% of our needs.

In 1975, the first full calendar year following the embargo, we consumed virtually the same amount of petroleum, but we imported 38% of our needs. In the period January-September 1976, imports rose to 40% of total consumption. All this happened in spite of Project Independence, in spite of patriotic appeals to the nation to save energy, in spite of mandatory speed limits on U.S. highways, and in spite of international agreements of various kinds between major oil-importing countries.

It is remarkable that the United States of America has been so totally inept in dealing with the energy problem. This suggests a fundamental misconception concerning the workings of the energy sector. Worse, this ineptness in dealing with a very small part of the U.S. economy (1.8% of its GNP, based on pre-embargo prices) raises serious questions concerning our government's ability to deal with the economy as a whole.

Despite numerous "statements" by politicians and others, the petroleum industry does respond to the most fundamental law of economics: the law of supply and demand. It is an indictment of our times that the supply and demand concept needs to be re-introduced and discussed. But, unfortunately, the supply and demand mechanism is thoroughly misunderstood today and many claims in support of it and against it are simply untrue, no matter how universally accepted.

A brief review of the supply and demand mechanism will set the stage for the ensuing discussions. For more detail, the reader is referred to Chapter 2, section *ii.*

The supply and demand mechanism expresses both production and consumption in terms of rates: so many tons of apples a year or so many millions of barrels of petroleum a day. This is of great importance when considering the problem of substitute sources of energy, such as coal. The *existence* of large coal reserves, important as it is, may assume a secondary role, and the really critical question is whether these reserves can be exploited at a rate sufficient to make up for declining petroleum production rates, and at what cost.

Everybody knows, of course, that demand curves are negatively sloped: For any given commodity, declining prices induce greater consumption rates. What is sometimes forgotten is the fact that demand for goods or services originates wholly and exclusively with the consumer of that commodity, reflecting the consumer's preferences and desires, but backed by purchasing power. No one, in a free market, can force the consumer to purchase any good that, at the going price, the consumer does not want. No corporation is big enough to brainwash the consumer or to force him to purchase goods he does not want, even though this is often asserted. As an example, Ford Motor Co. was unable to make the consumer buy the Edsel, even though it wanted the consumer very much to buy it. That failure again emphasized that *demand* for goods originates wholly and exclusively with the *consumer.*

The *supply* of goods and services originates wholly and exclusively with the *supplier,* and the supply curve is a reflection of the production rates that an industry is capable of maintaining at given prices. That curve is positively sloped: In a competitive environment, an increase in the price of a given commodity will invariably raise its rate of production.

The trigger mechanism is profits, and this is why a competitive market is such a remarkably efficient allocation mechanism: it reflects the ultimate vote of confidence of the consumer, who buys what he likes, if the price is right, and who in so doing rewards the producer who is most responsive to his needs, thus providing a premium for responsiveness.

Contrast that type of market mechanism with a centrally controlled market. The essential difference that emerges is lack of responsiveness in the latter, because rewards are not in tune with the consumer's vote.

It has been said that the consumer is king in competitive markets. This is a one-sided view, since it assigns the consumer a predominant role in the marketplace. The fact is that the producer and the consumer are equally important in a free market. Since demand originates exclusively with the consumer, and supply with the producer, some sort of accommodation must take place: The consumer cannot ignore the producer's production costs, nor can the producer ignore the consumer's preference.

Geometrically, an equilibrium exists where the demand and supply curves meet. Economically, what we mean is this: There is a price at which, in a competitive market, the consumption rate is exactly matched by the supplier's production rate. All the goods that are produced find a buyer in the market; there is no shortage and no surplus. This price is called the *equilibrium price* or market-clearing price, and it is stable. In the absence of cartel interference, by producers, by consumers, or by the government on behalf of either, the market-clearing price will prevail.

Suppose the government feels that a given industry deserves a "fair and equitable" rate of return and that it raises prices in that industry as it has, for example, in various agricultural commodities. As the price moves up from its equilibrium position, two things will happen: first, the consumer will react by consuming less of the now more expensive product; and second, the producer will react with higher production rates. Thus, the question arises concerning the ultimate outcome of the consumer's and the producer's conflicting reactions.

Because the demand for a good originates exclusively with the consumer, and because the producer cannot force his product on the consumer, it is the consumer's view that prevails in an overpriced but otherwise unaided market. No matter how much of a good is offered in the market, if it is priced too high, some of that commodity goes unsold and we have what is called a surplus. The term "surplus," as applied to any given good, is synonymous with the term "overpriced."

Similarly, if the government were to impose a price *below* the market-clearing level of a given commodity, a shortage will result. At unrealistically low prices, consumers would react by wanting to buy more, but marginal producers would be forced out of the market, with the result that overall production rates would decline. Unless suppliers were forced to produce at unattractive returns, the consumers' desires would go partially unfulfilled, and that, of course, is called a shortage. The important point here is that the term "shortage" is synonymous with "underpriced."

All of this is elementary economics and hardly needs to be explained, one would hope. But the fact is, while everybody has always known this, hardly anybody seems willing to apply it. For example, we all agree that this country is faced with an energy shortage, and in particular with an oil and gas shortage. But if the foregoing is correct, then it follows than an exactly equivalent method of expressing the problem is to say that oil and natural gas are underpriced in this country. We should ban use of the misleading terms "surplus" and "shortage"; their synonyms "overpriced goods" and "underpriced goods" are not only more accurate, they also suggest the remedy.

Underpriced Energy

Energy is underpriced in the United States. That is the problem. What has been done to solve it? For one thing, control of gas prices was taken from the market and given to the government as early as 1954, when the Supreme Court, in the notorious Phillips case, interpreted the Natural Gas Act of 1938 to also apply to interstate gas prices. The effect of this extension of governmental pricing powers was to substantially underprice natural gas, presumably in the interest of the consumer.

More recently, we have responded to rising world crude oil prices by refusing to let U.S. oil find its market-clearing level, and in so doing we are hurting the U.S. oil industry and playing into the hands of OPEC. The mechanism chosen was to establish a two-tier market for domestic oil, with both tiers below world prices. The fact that U.S. dependence on foreign oil grew during this period means that the two-tier price structure left U.S. oil underpriced, yet oil prices were rolled back even further in the Energy Policy and Conservation Act signed into law in December 1975.

Viewed from the price perspective, none of the measures taken make any sense, and the blame for our perverse response clearly is to be laid on the U.S. Congress. The Ford Administration, while initially unable to see through the problem, strongly supported and publicly advocated a return to decontrolled oil and gas prices. Of course, a presidential veto of the Energy Policy and Conservation Act would have been more consistent with the Administration's position, but 1976 was an election year and, as always, the political criterion prevails when it clashes with economic issues.

In addition to keeping energy underpriced in the United States, we have also managed to politicize the energy sector, and the oil industry in particular. It is a fact that prices in that sector are now set in accordance with prevailing political views, and these views are subject to capricious and unpredictable changes.

We know that U.S. oil supplies are declining, with or without help from Congress, and we know that oil and natural gas prices, if left undisturbed, will undergo a gradual increase with time, thereby stimulating the search for additional energy inside and outside the oil industry. Legislated prices, however, *cannot* be predicted.

The 1975 Energy Policy and Conservation Act might be replaced by a 1977 Act, or, for that matter, by a 1977½ Act. The resulting changes in the price of oil may be large or small, positive or negative—no one knows beforehand. And that makes planning impossible, and it therefore tends to discourage long-term capital investments in the energy sector.

For example, in January 1974, U.S. oil companies submitted bids on oil shales at the phenomenal rate of $41,319 per acre—expecting oil prices to rise in the future. They did not rise, and today oil companies have abandoned their oil shale ventures. Without question, the fundamental problem in the U.S. energy sector is underpriced energy, and especially underpriced gas and oil.

The Gravity of the Situation:
Some Questions and Answers

The public remains generally unaware of the serious nature of the U.S. energy situation, mainly because of the complexity of the subject and because the news media—source of most of the public's information—have done an extremely poor job of interpreting events related to energy supply, demand, and cost.

The questions that follow represent a cross section of the public's interest. They were submitted to the author by members of the American Business

Press, Inc. Some of the answers reflect the author's opinion, but they can form a basis on which interested readers may formulate their own replies.

What will be our energy situation 10-15 years from now?

Because energy has become a political issue, the answer to this question is speculative. As a general rule, the future energy situation will reflect our political will to accept the burden of developing new supplies of conventional and nonconventional energy. In uncontrolled markets that means higher energy prices, not just now, but prices that will continue to rise gradually in the future.

If we accept this burden, we will succeed, but not without a long-term, and perhaps permanent, loss in wealth. If we fail to accept this burden, there is no way we can succeed. There is no discernible concerted action in the U.S. to bear the cost of moving ahead.

What are the chances that the President and Congress will get together on a national energy policy?

They are, at present, remote. However, the 1976 election put a Democrat in the White House and a Democratic majority in Congress, so the President and Congress could just possibly be more finely attuned to each other in the coming term. However, the act of getting together on a national energy policy does not in itself resolve the issue: They might be getting together on the wrong policy, and if past performance outside the energy sector is a reliable guide, the chances of doing just that are not as remote as one might think.

When is Congress going to stop using energy as a football, and when, for once in its recent history, will it do something constructive?

This is a leading but legitimate question, and the answer is pessimistic. Congress may be expected to extend its heavy regulatory powers over more and more industries, to the detriment of the U.S. consumer. There have not been any recent serious free-market noises from Capitol Hill.

How much is the United States compounding its present and future energy problems by delaying a comprehensive energy plan?

The really critical problem is overplanning—no government plan at all would be the best plan. If this isn't apparent, imagine what a comprehensive government plan would do to *any* industry when prices are fixed and non-negotiable, while costs are rising; when allocation plans are drawn up for vital inputs; when you are asked to vertically and horizontally disintegrate, and when any performance deemed to be socially objectionable, such as above-average profits, opens you up to being called before a congressional committee where you will be abused with impunity.

In fact, much of the Administration's planning is needed not because the U.S. energy industry is incapable of doing its own planning, but because a capricious Congress all but renders private planning impossible.

The essential corporate planning function today is not how to react to changing market conditions, but how to anticipate the political mood of the U.S. Congress and, having done so with more or less success, how to take a reasonable defensive posture.

How long can petroleum be counted on as a basic energy source?

Hydrocarbons are an exhaustible resource, and sooner or later the supply will run out no matter what policy is pursued. The usual measure of a country's current resource availability is the so-called reserve-life index which, in the case of crude oil, gives the number of years the oil will last if only proved resources are considered; if there are no oil imports; if there is no switching to other forms of energy; if there are no further additions to existing reserves through exploration; and if oil can be produced from wells at top capacity, until the well is drained.

Each of the preceding criteria is subject to challenge. There is, first of all, the great controversy of what does or does not constitute proved reserves. Moreover, there are oil imports, and the greater these are in relation to our crude oil consumption, the more the United States is exposed to a very sudden shock should developments in the Middle East produce another embargo.

In addition to currently proved oil and gas reserves, there are two more sources of additional hydrocarbons available to the U.S. economy: increased recoveries from existing fields, including fields that have been exhausted through conventional recovery methods, and newly developed reserves through intensive and more sophisticated exploration methods.

As was shown in Chapter 7, the current U.S. reserve-life index stands at 5.5 years. Secondary recovery, applied to all known reserves, will raise the index to 11.5 years. Tertiary recovery in the opinion of some industry authorities, could raise the index to 19 years (others are not so optimistic). Exploitation of heavy crude oils (API gravity less than 25°) will add another five years to the index, for a total of 24 years.

Shale oil is, and will remain for a long time, a white elephant. It is always going to be expensive, perhaps too expensive. In any event, it does not add hundreds of years to our energy balance—it adds perhaps 40 years, if we can ever produce it commercially. In addition to the expense of producing it, there is the problem of building it up to an acceptable rate, and the odds are overwhelming against this happening.

Tar sands are a pipe dream—we don't have any of consequence. The only sizable North American tar sand deposits are in Canada and, therefore, outside of our jurisdiction. Of course, Canada is not likely to join OPEC, but because the Canadians are becoming very apprehensive about their own energy supplies, they are already phasing out crude exports to the United States.

The one major U.S. fossil fuel outside of hydrocarbons is coal, but even there our expectations are probably too high. At 1970 production rates, U.S. coal reserves are good for 300 years. At a 3% growth rate, the reserve-life is only 85 years; at a 5% growth rate, it shrinks to 60 years. And, of course, such a growth rate becomes imperative if we switch to coal in our search for alternative energy—imperative, but not necessarily technically feasible, because of the ever-present capacity problem.

What about transition time? In a free market it will be gradual. Remember, prices will rise gradually, so the reallocation process will be slow. In addition, oil fields don't play out overnight; their productive capacity declines with time. For this reason, a reserve-life index of 5.5 years does not mean that we have sufficient amounts for that long and nothing thereafter. People sometimes read that into the index, and their reaction is panic or disbelief, maybe both. Even if we add no new oil to our current reserves, there would still be producing oil wells in this country 30 to 40 years from now. The East Texas Field, admittedly a giant, has been on production for more than 40 years now, and the end is not yet in sight.

And then, of course, drilling and exploration activities have not stopped. On the contrary, we now have almost twice as many rotary rigs running as we had in 1971, and we have vastly more sophisticated oil exploration methods than we had a few years ago. Still, oil fields are getting harder to find: the obvious ones have already been found, and a sustained finding rate of crude oil reserves requires more and more effort, and money.

What about the effect on life in the United States? The impact on life is often drastically exaggerated. It isn't as though energy will cease to be available at any cost. It will simply be scarcer and, therefore, more expensive. That does not mean a return to the cave. For example, Germany produces almost as high a per-capita GNP as the United States, with less than one-half the per-capita energy consumption. Why? Because the Germans work harder, but also because energy costs are some two times as high there as they are here. At those prices, the Germans are naturally more energy conscious than we are.

Is there any alternative to higher prices as a means of conserving energy?

Americans have become spoiled and the adjustment to a more rigorous life is painful. The volunteer conservation program is not working for a simple reason; a slight misunderstanding let us say: Everybody is volunteering his neighbor to save energy, but not himself. Volunteer programs of any kind make good rhetoric but poor economics.

A country cannot be run for long by appealing to the patriotic feelings of its citizens. It has been tried before, without success, in the areas of spending, saving, foreign investments, to name a few. It has always failed. Indeed, economic theory says that it is destined to fail.

The truth is, higher prices are the only long-term conservation tools that will work, and we might as well get used to it. Moreover, higher prices will redirect our research efforts into what are at present prohibitively expensive alternative energy sources such as syncrude or gas from coal, hydrogen through nuclear plants, etc.

Is the public willing to pay the money price and the environmental price to develop energy resources?

Intensified energy production is necessary, but this can contribute to polluted cities and to the creation of a hostile environment. Clearly, controls are needed, since the free market has failed here. The consumer will buy the cheapest of two products, not the one that was made using environmentally preferable production methods. Hence the urge to externalize production costs by belching smoke out of chimneys, or by putting pollutants into streams.

However, it is not ultimately a matter of paying a money price. It's a matter of giving up some of the things we are now enjoying so that we can once again enjoy clean air and clean water. Nor is it a matter of cleaning up once and being done with it forever. This is a rate problem, too, and it requires a permanent reallocation of resources for an ongoing effort to clean up the environment and to keep it clean.

Solar energy is low-density energy. It won't make much of a dent in U.S. energy. Wind, tidal, and geothermal energy also are of limited importance in the overall energy situation. In the future, the pressures of declining energy reserves will curb some of the currently exaggerated environmental constraints.

What energy source comes after oil and gas?

There are no two ways about it, the ultimate fuel will have to be nuclear, for the generation of electricity and, more importantly, for the generation of hydrogen from sea water, that inexhaustible and environmentally clean-burning propellant fuel that will keep our cars running and our planes in the air.

All other energy sources have at least one of two problems, and often both: they are exhaustible or they cannot be delivered in sufficient quantities.

Methanol from wood can power vehicles—as it did in Germany during World War II—but not enough can be produced to power the 50-60 million cars now on the road, not counting trucks and railroads.

The use of natural gas on the East Coast in lieu of fuel oil is possible, but not likely. Plans are under consideration to connect the East Coast to Arctic gas reserves, but at best Arctic gas availability is years away from reality. And it will be expensive: $2.50 to $3 per Mcf in 1975 dollars, or 50% more than the present cost. The other alternative, imported LNG, is not going to be any cheaper.

Beyond energy, what are the most pressing problems in the United States?

As mentioned earlier, the U.S. government's ineptness in dealing with the energy crisis is symptomatic of its general ineptness in dealing with the U.S. economy, which is at present in a very precarious position. We have not been able to cure the stagflation problem for six years now. In fact, in 1969-1970, we had an unemployment rate of 6% and an inflation rate of 6%, and we panicked and subjected the nation to wage and price controls. Remedial policies were needed in 1971, but we picked the wrong policies. For reasons that would take us too far afield, we have resorted to accommodating economic policies ever since we have had economic problems. We are masters in the art of buying time.

There are many symptoms of disease in the nation. New York is not unique, it is only first. We have been printing money like it is going out of style, and as a result, it is. For example, we nearly doubled the money supply in the last 10 years (from $166 billion in December 1965 to $295 billion in December 1975), while the real GNP rose by only 28%.

Most do not realize that money printing has been so prodigal in the United States. Hardly anyone ever takes a long-term look at money. The papers are full of speculation concerning this or that shift in short-term monetary policies, but there are no long-term assessments.

The same holds true for deficit spending: In fiscal 1976 alone, the U.S. added another 20% to its total national debt, as accumulated over the past 200 years. We are operating on a precarious line of credit, and that line cannot hold forever.

The government, in financing its fiscal 1976 deficit of $65.6 billion, went to the credit markets and asked for new loans to the tune of *$260 million every working day.* In the process, it put upward pressure on the interest rate and displaced the private sector from the credit markets. The result is reduced net investment in all industries, including the energy industry, whose capital requirements are now larger than ever. The result is also a shrinking private sector that will be harder and harder pressed to support a growing public sector.

There are many more symptoms of a slow erosion of our economic strength, so slow that they remain unnoticed. For example, in 1950, government transfer payments to persons, which are essentially payments for nonproduction, ran at $14.4 billion a year and were equal to national defense outlays. Today, transfer payments amount to $175 billion, double our defense effort.

Our entire economic thinking has become warped. Trapped in the Keynesian dogma, we believe spending, and in particular government spending, is the answer to our problem, and nobody seems too worried about production. Illusions have become articles of faith in economic policy planning: the illusion that a full-employment budget is noninflationary, regardless of its deficit; the illusion that a trade-off exists between inflation and unemployment; the illusion that cost-push forces are self-limiting and that cost-of-living escalators are noninflationary.

The long-term outlook for the U.S. economy is bleak, but not because of the energy crisis. And neither will the energy producers (OPEC) bring about our demise. We need no outside help, since we are expert at digging our own graves. This will be discussed in detail next.

iii. Can the U.S. Economy Collapse?

In 1965, Moody's Investor Service, Inc., downgraded New York City's municipal bonds from *A* to *Baa,* not without strenuous objections and some verbal abuse by the city administrators. After all, *Baa* is the lowest nonspeculative bond category there is. At one step below, *Ba,* the bonds are considered so risky that their purchase is not recommended, and, indeed, their interest rates are no longer listed: the bonds are considered too speculative for investment purposes.

Today, some 12 years after Moody's downgrading of New York City's municipal bonds, the financial collapse of the nation's largest city is imminent; it would be history today, had not New York State and the federal government temporarily bailed out the city, wisely or not.

The fundamental reason for New York City's problem is the same perennial cause of any financial collapse: expenditures exceed income. If and when that happens over extended periods of time, any economic entity, private or public, must fold, including the U.S. government.

The average informed American today believes that the U.S. government cannot go broke. That's part of the doctrine that has been taught to at least two generations of college graduates. This doctrine, which is in fact a very superficial interpretation of Keynes, holds that the national debt is no burden to a country; that it is a debt we owe to ourselves, unlike any other debt; and that is like replacing $10 in our right hip pocket with an IOU and putting the money in the left hip pocket.

This message is still relayed to the current generation of college students, as a spot survey of the most popular introductory economics texts reveals. Paul A. Samuelson, Nobel Laureate in Economics and writer of the most successful economics text in history rejects as "misleading" the notion that the public debt is "like a load on each man's back." C.R. McConnell states that "the public debt is one we owe to one another." Only when the public debt is held by foreigners would it be like a private debt, according to McConnell. On the question as to whether a big public debt can bankrupt the economy, G.L. Bach, in another monumental economics text, says, ". . . extremely unlikely, so long as the debt is domestically held." There is no burden, according to Bach, since "large annual interest costs . . . are only transfer payments within the economy—from one of our pockets to the other."

Are there no exceptions from this complacently self-defeating view of the national debt? There are, but they are just that—exceptions. For example, A.A. Alchian and W.R. Allen have challenged many accepted views on contemporary economics in their introductory text, *University Economics*. "The situation is really not much different from that for private debt," and, "not much advance in analysis or understanding is evoked by the saying 'we owe it to ourselves.' " Samuelson, McConnell, Bach, and countless other economists have embraced the conventional wisdom, the economic dogma of the West: a rigorous adherence to a narrow interpretation of the writings of an early twentieth century economist, J.M. Keynes.

In addition to the pernicious notion that the national debt entails no burden, the dogma has produced other fallacies. For example, under this doctrine it is generally held that a full-employment budget is noninflationary; hence this argument opens the door to unchecked governmental deficit spending. Then there is the Phillips Curve and the related doctrine that there exists an inevitable trade-off between unemployment and inflation. Who, in America, does not believe that? Yet, the concept is unfounded on both theoretical and empirical grounds.

A third fallacy holds that cost-push forces are self-limiting, that cartel forces (including labor cartels) can drive up prices only so far, and no more. A fourth and related fallacy holds that cost-of-living escalators are noninflationary.

All of these fallacies are part and parcel of our official economic thinking. Public policy is made on the basis of these and other fallacies, and they are extolled to the current generation of future leaders as principles on which to base future policy decisions.

Obviously, if the national debt represents a burden to the U.S. economy, then it must be true that this burden has the potential of becoming too much for the economy to bear, that a point might be reached where the economy simply collapses under its burden, similar to the way New York City is collapsing now. That point exists and it can be defined. What's more, calculations can be made to determine with reasonable accuracy when that point will be reached.

For an individual, Doomsday is the day on which he can no longer meet his payments—when his interest obligations exceed his capacity to borrow. Unable to borrow from Peter to pay Paul, he knows that Paul will foreclose on him. How does an individual get into this predicament? By living ahead of himself, that's how; by developing a life style he simply cannot afford; by basking in luxury that is not his.

If an individual uses borrowed funds to expand his productive potential, he will not necessarily face bankruptcy. The business he is financing with that money might produce the revenues and, hopefully, profits with which to pay off his debt and reduce his interest payments. But suppose he uses his credit to build a large home instead, and a lakeside cottage, with a sailboat and a motor yacht. All of these luxury goods need to be maintained, and the maintenance financed through additional loans—until none can be had.

Bankruptcy will result when the maintenance of an existing system can no longer be financed through profits or loans. This holds true for any system, including the productive one: the business or manufacturing operation. But in the case of a nonproductive operation, if it follows a path of irresponsible and irreversible growth, its date with fate is a foregone conclusion. There is *never* any hope, even in its incipient stage, for its ultimate survival.

This holds true for nations as well. The U.S. government could conceivably displace the private investor from the capital markets. If this happens, and if a concurrent inflation rate of 10% or more renders depreciation more and more deficient as a mechanism to maintain private investment at its current level, then the government will be the ultimate owner of all productive resources. This would have serious repercussions on U.S. living standards in view of the well-known inefficiencies of centrally planned economies. More importantly, it would drastically curtail freedom, with the government making decisions concerning every facet of private life, but there would at least be a productive process. If, on the other hand, the government were to use its credit for the extension of existing luxury goods and the building of new ones, the picture would change—this is no longer the replacement of an efficient productive process with a less efficient one. This is the abolishment of production. And that process is as doomed, from the start, as is the individual's unchecked acquisition of luxury goods through loans.

Governmental expenditures for nonproductive activities are called transfer payments. Medicare, unemployment compensation, the food stamp program (which has recently been extended to include people on strike, i.e., who refuse of their own free will to work—that's not nonproductive, that's downright counterproductive), school aid including free breakfast and lunch, are all such payments. Collectively, they currently run at $175 billion a year. That's about 10% of the U.S. GNP, up from 5% in 1960.

Whether government transfer payments are justifiable on humanitarian grounds is not the issue here. The question is, simply, whether the United States can afford them and, if not, what the consequences are of having them anyway.

Bankruptcy is inevitable when a system what was built up through loans can no longer be maintained through loans. For a government, or any economic entity, this means that bankruptcy occurs when interest payments on the debt exceed the unit's borrowing capacity.

What does this imply for a nation? A nation can and will go bankrupt when the interest payments on its public debt exceed the net capital generation of that nation. When this happens, the collapse of the economy is imminent.

Of course, this will not necessarily take the form of an abrupt change, although it might. The market may become increasingly apprehensive as Doomsday approaches, as it did in New York, until the inevitable is there for even the most insistent optimist to see.

Doomsday, then, marks the beginning of an economic collapse. But what, precisely, is meant by the term "economic collapse"? This term is here defined as a combination of unemployment and inflation so rampant that the market ceases to function effectively. An admittedly arbitrary rule is that when the sum of percentage inflation plus two times percentage unemployment equals 50 or more, that is a collapse of the free-market system as it now operates in the United States.

For example, 25% unemployment and no inflation means collapse under this definition. Zero unemployment and 50% inflation also means collapse, as does 10% unemployment and 30% inflation. Other collapse combinations under the terms of the preceding definition are listed below.

Economic Collapse Defined

Unemployment %	Annual inflation %
0	50
2.5	45
5.0	40
7.5	35
10.0	30
12.5	25
15.0	20
17.5	15
20.0	10
22.5	5
25.0	0

By this definition, the Great Depression constituted an economic collapse, with 25% unemployment and no inflation (in fact, declining price levels). That's why 25% unemployment was chosen as the one extreme of the bi-polar unemployment-inflation continuum. The other pole, 50% inflation, is based on the fact that public confidence in government-issued fiat money tends to break down somewhere near that threshold, and barter begins to replace the money economy.

The U.S. economy began to display signs of distrust of its fiat money in July-August 1974, when the consumer price index rose from 148 to 149.9, or at an annual rate of 15.4%. At that time many U.S. corporations were experiencing difficulties in obtaining firm price quotations for future deliveries, and barter deals were openly discussed in the market.

In trying to pin down Doomsday, two factors need to be considered: the net capital-generating ability of U.S. economy and future interest payments on the national debt. To get the latter factor, projections of future deficit spending and the resulting increase in national debt, as well as of applicable interest rates, are required.

Annual U.S. deficit spending from 1947 to 1975, plus the budgeted deficit spending level for fiscal 1976, is plotted in Figure 9-1. The latter magnitude of $60 billion is known to be conservative, since the actual budget deficit in 1976 turned out to be $65.6 billion. Two statements can safely be made in observing the deficit spending curve: first, it is erratic, with violent year-to-year fluctuations; and second, its overall trend is up. Because of its volatility, the deficit spending curve does not lend itself to future projection. Here the well-known technique of moving averages comes into play.

The second curve in Figure 9-1 represents the five-year moving average, with time, of U.S. deficit spending. For example, the first point in 1949 represents one-fifth of the total deficit spending in 1947, 1948, 1949, 1950 and 1951. The point is negative, which means that the U.S. government had a budgetary surplus (a negative deficit) over the five-year period.

The second point (1950) represents a similar average, covering the period 1948 through 1952, and so on. The overall curve is much smoother than the year-to-year curve, since the averaging procedure tends to pull in high and low extremes. That five-year moving average curve can now be projected into the future.

In making that projection it is assumed implicitly that the forces currently at work in pushing up governmental deficit spending, whether they are identified or not, will continue to do so in the future.

Is this a realistic assumption, and if so, why are those forces not restrained? Somebody should ask the administrators of New York City that question. Chances are, they have the answer. It will be a political answer, of course, which is not the subject of this discussion. Suffice it to say here that political considerations will act as a blocking agent in curbing these forces, and deficit spending will follow the approximate projection path of the five-year moving average curve of Figure 9-1, barring a complete reversal of the government's current attitudes.

The projection of future deficit spending is shown in Figure 9-2. Using that curve as the basis of projected future deficit spending, their estimated magnitudes for 1975 through 1990 are listed in Table 9-2. Again, as can be seen in column (2), the annual deficit spending rate is not projected to reach $65 billion until 1980, when in fact it did so in 1976. If anything, Doomsday is much closer at hand than the ensuing calculation suggests, and that is well worth keeping in mind.

Column (3) of Table 9-2 shows the national debt as of 1974 and where it will be in future years. The 1975 figure of $519 billion is obtained by adding that year's deficit to the 1974 debt. Given the national debt as listed in column (3), the annual service charge on that debt varies with the rate of interest. Columns (4) through (7) give the annual service charge for each year's

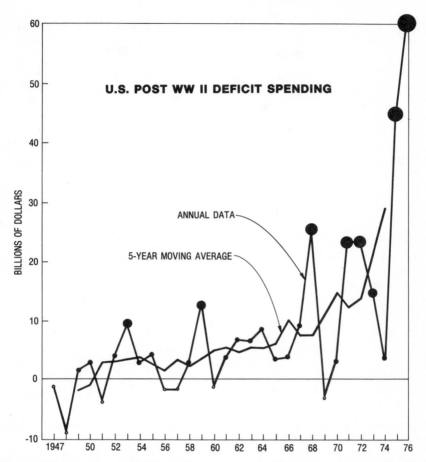

Figure 9-1. *Annual U.S. deficit spending from 1947 to 1975, plus the budgeted deficit spending level for fiscal 1976.*

national debt for interest rates of 7%, 9%, 11% and 13%, respectively. For example, given a national debt of $798 billion in 1980, the annual service charge on that debt will be $56 billion if the interest rate is $7%, $72 billion if the rate is 9%, and so on.

But what will the interest rate be in future years? One thing is sure: it will go up, if left undisturbed. The problem with predicting interest rates is that they are easily and frequently manipulated through monetary policies. Still, at the current rate of deficit spending to the tune of $65 billion, the government is entering the financial markets with requests for new loans corresponding to no less than $260 million every working day, based on 250

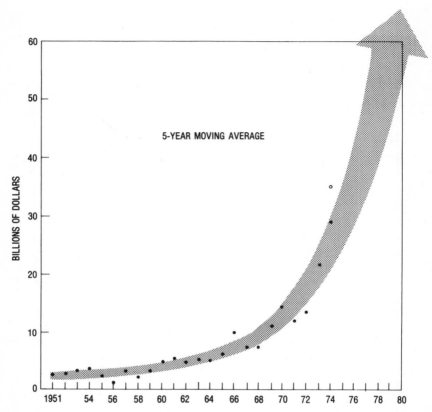

Figure 9-2. *U.S. post-World War II deficit spending and projected future deficit spending (five-year moving average).*

working days per year. That kind of a massive credit demand will find the business sector with a new and substantial competitor. Between the business sector and the government, scarce credits will be bid up in yield: The interest rate cannot help but rise in the long run, regardless of intervening short-term monetary policies.

Indeed, easy-money policies are self-defeating in reducing interest rates: Easy money will reduce interest rates in the short run, but it will also fuel the flames of inflation, and rising prices will ultimately raise the rate of interest above the pre-easy-money level. For example, if the interest rate in an inflation-free economy is 5%, then it stands to reason that it should rise to 15% when the inflation rate rises to 10%, because lenders will not voluntarily transfer a part of their wealth to borrowers by agreeing to be paid back in soft money, without compensation for inflation.

Table 9-2
Projected Deficit Spending, 1977-1990
(Billions of dollars)

End of Fiscal Year[1]	Projected Annual Deficit	National Debt	Annual Service Charge on National Debt			
			r = 7%	r = 9%	r = 11%	r = 13%
(1)	(2)	(3)	(4)	(5)	(6)	(7)
1974	—	475	33	43	52	62
1975	44.2[2]	519	36	47	57	67
1976	60.0[3]	579	**40**	52	64	75
1977	41.0	620	43	56	68	81
1978	49.0	669	47	60	74	87
1979	59.0	728	50	66	80	95
1980	70.0	798	56	72	88	104
1981	80.0	878	61	**79**	97	114
1982	90.0	968	68	87	106	126
1983	100.0	1,068	75	96	117	139
1984	110.0	1,178	82	106	130	153
1985	120.0	1,298	91	117	143	169
1986	130.0	1,428	100	129	**157**	186
1987	140.0	1,568	110	141	172	204
1988	150.0	1,718	120	155	189	223
1989	160.0	1,878	131	169	206	244
1990	170.0	2,048	143	183	225	266

[1]June 30 of year shown
[2]Actual
[3]Budgeted

Thus, the rate of interest is sure to rise with the government engaging in peacetime deficit spending at a level considerably higher than the deficit spending level during World War II. How fast will it rise? It is assumed here that the rate of interest will rise by 2% every five years. This is probably a conservative estimate if the rate of inflation is held at about 10% over the coming years through "appropriate" easy-money policies.

In any event, the 10% assumption on inflation is not terribly important to the outcome of the calculation concerning the collapse of the U.S. economy. If the inflation rate is allowed to go higher, accelerating with time, the interest rate will not rise as fast. This would mean that a portion of the service charge on the national debt is shifted to the U.S. public via rising prices instead of rising direct interest payments, but payment *will* be made, and made by the U.S. public. In the meantime, collapse still approaches as the public's confidence in the government's money declines. The following calculation, then, will assume a 2% increase in the rate of interest every five years.

Starting with a 7% average interest rate on credit instruments issued by the government, the 1976 service charge on the national debt will be about $40 billion, as indicated in column (4) of Table 9-2. In 1981, the service charge will be subject to an interest rate of 9%, and it will amount to approximately $79 billion [column (5)]. The annual service charge on the national debt in 1986 will rise to $157 billion [column (6)].

The growth of the service charge with time is shown in Figure 9-3, where the various curves represent the respective charges for given interest rates, corresponding to the magnitudes in Table 9-2. The dotted line in Figure 9-3 reflects the service charge, taking into consideration the gradual increase in the rate of interest. For example, Point A in Figure 9-3 ($40 billion) corresponds to the magnitude in column (4) of Table 9-2 and represents the pro-

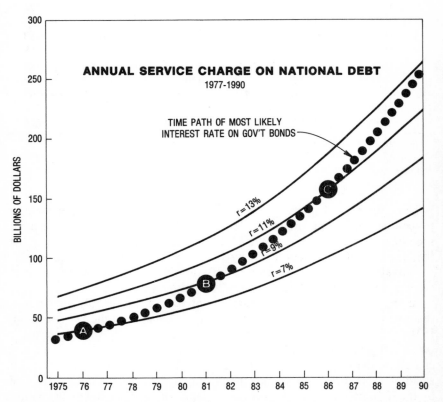

Figure 9-3. *The annual service charge on the national debt (1975-1990).*

Table 9-3
The National Debt and Its Cost
(Billions of dollars)

Year	National Debt	Service Charge
(1)	(2)	(3)
1976	579	40
1977	620	45
1978	669	51
1979	728	59
1980	798	68
1981	878	79
1982	968	92
1983	1068	106
1984	1178	121
1985	1298	138
1986	1428	157
1987	1568	179

jected service charge for 1976. Table 9-3 summarizes the national debt and its service charge over the next 12 years.

If the current trend in deficit spending continues, column (3) of Table 9-3 shows the annual service charge of the U.S. national debt in the period 1976 to 1987. To say that this service charge will rise would be an understatement. Given a near-20% increase in the national debt in fiscal 1976 alone and accelerated growth thereafter, plus rising interest rates, the projected increase in the service charge by a factor of 3 between 1976 and 1984 is not at all unreasonable—ominous, yes; unreasonable, no.

The government, by definition, finances its budgetary deficit by taking loans from the public. Of course, there is a ceiling on the amount of credit the public is willing to provide each year, and that ceiling is the net capital-generating capacity of the U.S. economy.

People in modern society have to do at least two things: eat and pay taxes. For the economy as a whole, funds that are left over after consumer expenditures and taxes can be used to finance the production of capital goods. That is called gross investment in the United States GNP accounts, but it does not reflect the net capital-generating capacity of a nation, simply because existing capital goods wear out and have to be replaced. A good deal—in fact more than half—of gross investment represents replacement capital goods, which are funded through so-called capital consumption allowances. Deducting these allowances from gross investment leaves net investment: This is the net addition to a nation's capital stock, and it represents that nation's net capital-generating capacity, provided that its government has a balanced budget.

Net investment, then, represents funds that the people in a nation have set aside for future growth. Its source is the people's total after-tax income—what's left of it after consumption. In that sense, saving is nonconsumption and, again, with a balanced budget, is equal to net investment.

If the government does not balance its budget, the economy finds itself with two claimants on the people's savings stream: the business sector for net addition to capital stock, and the governments, federal, state and local, to cover their deficits.

Even though state and local governmental deficits are sizable, nearly $14 billion in 1973, they will be ignored here simply because the most pernicious, i.e., nonproductive, activities are funded by the federal government. This omission is conservative, since the inclusion of state and local deficits would tend to move up Doomsday (cf. New York City).

Of course, the federal government has not balanced its budgets lately. By dipping into the people's savings, it has mopped up a part of this nation's growth potential. For example, Table 9-4 shows the U.S. net capital-generating capacity for the fiscal years 1972 through 1975, and how it was used.

As the table shows, the U.S. net capital-generating capacity amounts to approximately $100 billion, not counting state and local deficit spending. Adding on the latter, the net capital-generating capacity is approximately equal to $110 billion, give or take a few billion dollars.

The service charge on the U.S. national debt will be $110 billion in 1983-1984. At that time the United States will find itself in the same position that New York is in today. For the nation as a whole, that will be the day when private investment is limited to replacing worn-out capital goods, and barely

Table 9-4
Net Capital-Generating Capacity
U.S. Gov't Deficits
(Billions of current dollars)

Calendar Year	Net Capital-Generating Capacity	Used for Net investment		Used for deficit spending	
		Amount	%	Amount	%
1972	100	83	83	17	17
1973	111	103	93	8	7
1974	89	78	88	11	12
1975	106	31	29	75	71

Source: Federal Reserve Bulletins, April 1976 and May 1974, Financial and Business Statistics Sections.

even that, since the ongoing inflation rate will render capital consumption allowances insufficient for that purpose. Then what? There are several alternatives, none good.

The United States can resort to printing money. If the Federal Reserve System resists, pressure will build up in Congress that may well lead to a federalization of the Central Bank which would promptly become a political football, willing to print money at the behest of Congress. The result would be hyperinflation and a collapse of the economic system shortly thereafter. Current indications are that the Federal Reserve System and the U.S. Congress are headed toward a clash that could soon end up placing the system under political controls.

Or the United States can cancel its national debt. Surely, this would not be done overtly. It would probably take the form of a "freezing" of interest payments. That, too, would misfire, since the U.S. government would in the process destroy whatever investor confidence there is at that time. After the first such freeze it seems unlikely that the government could sell additional bonds.

If, however, an outright *cancellation* took place, it would become immediately apparent that this is not an act of returning from one hip pocket what used to be in the other. It would immediately be clear that we are not dealing with a debt that "we owe to one another." To see this, a look at who owns these government bonds is warranted.

In December 1975, $91 billion in public debt issues were held by individuals, most of them in the form of savings bonds. Cancelling that part of the national debt means that the life's savings of many people would be wiped out by the stroke of a pen. Another $67 billion in public debt issues consists of foreign and international accounts in the United States. International confidence in the soundness of the U.S. economy would crumble if that part of the debt were cancelled.

The potentially most damaging effect would be in the commercial banking sector. Banks hold some $86 billion in public debt issues. Many banks would collapse if that part of the debt were cancelled, and in the process they would drag down countless corporations, noncorporate businesses, and individual savers. Banks are all the more vulnerable, since they are prohibited from holding equity instruments in their own right.

State and local governments hold substantial amounts of public debt issues, nearly $34 billion of them. They, too, would get into financial difficulties should the national debt be cancelled.

Other institutions jeopardized by their U.S. bond holdings are corporations ($20 billion), insurance companies ($10 billion), mutual savings banks ($5 billion), and various others. Also included are savings and loan associations, non-profit institutions, corporate pension trust funds, etc., to the tune of $38 billion.

Decidedly, the U.S. economic system cannot survive in its present form if the national debt is cancelled outright or if interest payments are frozen. But what about taxation?

The problem here is that a shrinking private sector is burdened with an expanding nonproductive sector. If the government were to confiscate via taxation what it cannot borrow, the result would be the same as if it went the loan route. By 1983-1984, the government will still have used all available funds, and Doomsday will still be at hand. Of course, the national debt would be relatively smaller then, and with it the service charge on the national debt. But that reduced service charge would have to be met out of greatly reduced after-tax incomes, and the problem would in essence remain the same.

There is no other way: substituting ever-growing nonproductive services for productive activities is a form of slow but inexorable self-destruction.

There are many factors that could change the timing of the impending economic collapse. Without doubt, the most important of these will be the spending pattern of the new Carter Administration. Keeping in mind that government spending is, for the most part, mandated by legislation, there are only two ways the current trend can be reversed: The President must either convince the 95th Congress that its spending attitude is self-defeating, or he can continue President Ford's more or less successful policy of vetoing spending measures.

Time is running out. To the extent that inflation has rendered capital consumption allowances partially ineffective as a mechanism to perpetuate the nation's capital stock, we are already eating into our capital base without realizing it. This, by the way, holds true also for profits that are fictitious, since their calculation assumes no inflation.

If an economic collapse is within the realm of possibility, where does this leave the corporation or the individual? It is clear that corporate policies would have to be totally reoriented: survival under adverse conditions would become the major policy objective, through production shifts to basic necessities and away from luxury goods, coupled with portfolio shifts out of money and credit instruments and into real assets or options for real assets. The establishment of corporate liability barriers by voluntary corporate disintegration may become a popular defensive strategy.

The average wage-earning individual has fewer options. He may go into debt, but his chances of paying off a substantial debt on real assets, using worthless money, are wiped out if he loses his job before the onset of hyperinflation. Neither credit instruments nor equities offer much safety. His portfolio would have to be diversified to spread the risk. Some gold should be held, since it cannot be debased; certainly, some real estate, especially if his state laws admit declaring his house a homestead—out of reach of the creditors. He must hold absolutely no money. If credit instruments are held, these must be disposed of rapidly once a predetermined trigger inflation rate is reached, say 20%. If equities are held, these should be in solid corporations not subject to the consumer's whim: bakeries, shipping, agricultural goods, extractive industries—especially oil companies.

In any event, a financial collapse of the United States is possible. Worse, it is inevitable unless current economic policies are drastically revised. Total U.S. government commitments are much larger than the national debt shown here. For example, social security obligations currently amount to nearly a trillion dollars, and it is no secret that the social security reserve fund will soon run dry.

It is clear that OPEC had nothing to do with our financial commitments, but OPEC *will* be blamed for the U.S. collapse, which will create shock waves worldwide, wreaking economic chaos on many other nations. This imposes a solid defensive strategy on OPEC now; a strategy that includes close observation and analysis of economic and political events in the United States and in other oil-importing countries.

As we have seen, it is possible to avert the collapse, but this would require that our government, at every level, put economic considerations ahead of political objectives. Whether this is likely to happen in a "democratic" society as we have it is for the reader to judge. We shall all know soon enough.

Index

A

Abu Dhabi, 96-97, 105
Aggregate demand and supply
  money, and inflation, 13, 42-43
Alaskan oil, 180, 184
Algeria, 124
Alternative energy sources (*see* Energy sources)
Aluminum, 46
Arab oil embargo, 8-9, 57, 60, 76, 137, 181, 184
Arab oil-exporting nations (*see also,* OPEC)
  import needs, 198
  perceptions of U.S. economy, 194-199
Atomic Energy Commission (AEC), 159-160, 168, 173-174
Australia, 46, 56-59, 134
Austria, 56-59, 134

B

Bahrain, 96-97
Balance-of-payments deficits (*see also,* Deficit spending)
  defined, 10
  and fixed exchange rates, 77-81
  and international inflation, 11-19

Barre, Raymond, 13
Base period indexation, 118-120
Belgium, 76, 134
BOE (barrel of oil equivalent), 4
Boiling water reactor, 160-165
Bolivia, 80
Bonds (U.S.), 106, 133, 219-20
Breeder reactors, 168-173
Bretton-Woods Agreement, 10-12, 75-78, 81
British system
      vs. metric system, 3
Buffer stock borrowing, 80

C

Cabinet Task Force on Oil Import Control, 18, 40
Canada, 46, 76, 134
Capital
    in the petroleum industry, 20-21
    preservation of, 107-110
    U.S. capacity to generate, 218-219
Cartels, 38-39, 42-43, 134-135
Chemical energy, 3-4
Chile, 46, 82-83
Clean Air Amendments, 148, 180
Coal, 136, 147-149, 180-181
Committee of Twenty, 76, 83-84
Compensatory borrowing, 79-80